Translations of Mathematical Monographs *Volume 6*

HARMONIC ANALYSIS
of Functions of Several Complex Variables in the
CLASSICAL DOMAINS

by

L. K. Hua

AMERICAN MATHEMATICAL SOCIETY
PROVIDENCE, RHODE ISLAND
1963

ГАРМОНИЧЕСКИЙ АНАЛИЗ
ФУНКЦИЙ МНОГИХ КОМПЛЕКСНЫХ ПЕРЕМЕННЫХ
В КЛАССИЧЕСКИХ ОБЛАСТЯХ

ХУА ЛО-КЕН

Издательство Иностранной Литературы

Москва 1959

Original Chinese text published by Science Press, Peking, 1958

Translated from the Russian
by Leo Ebner and Adam Korányi

Publication aided by grant NSF-GN 57
from the
NATIONAL SCIENCE FOUNDATION

Text composed on Photon, partly subsidized by NSF Grant G21913

Library of Congress Card Number 63-16769

Standard Book Number 821-81556-3

Second Printing, 1969

Printed in the United States of America

CONTENTS

EDITOR'S FOREWORD

The subject of the present monograph by the well-known Chinese mathematician L. K. Hua is the study of the important classes of bounded domains in the space of several complex variables. These domains were first described by E. Cartan. Their study is very important in the representation theory of Lie groups, in the theory of homogeneous spaces and in the theory of automorphic functions of several complex variables. Consequently, the book of L. K. Hua has close connections with the works of other authors in the above mentioned fields. We are thinking here of the works of E. Cartan, I. M. Gel′fand, F. A. Berezin, R. Godement and Harish-Chandra on the theory of spherical functions on Lie groups, numerous works on the finite-dimensional representations of Lie groups, the works of I. M. Gel′fand, M. A. Naĭmark, F. A. Berezin, M. I. Graev, Harish-Chandra on infinite-dimensional representations of Lie groups, and the results of the theory of automorphic functions of several complex variables, some of which are to be found in the book of C. L. Siegel, *Analytic functions of several complex variables,* Institute for Advanced Study, Princeton, N. J., 1950; Russian Translation, IL 1954. We should also mention the results of I. M. Gel′fand and M. I. Graev [**5**] on the representations of Lie groups on homogeneous spaces; the very recent interesting results of F. I. Karpelevič [**10**] and I. I. Pjateckiĭ-Šapiro [**11**] on the boundaries of symmetric spaces, which were applied by these authors to the theory of harmonic functions on symmetric spaces and to the theory of automorphic functions of several complex variables; the results of I. M. Gel′fand and I. I. Pjateckiĭ-Šapiro [**7**] on representation theory and the theory of automorphic functions (see the supplementary bibliography at the end of the book). This highly incomplete list already shows that the book of L. K. Hua is closely related to some of the most rapidly developing branches in contemporary mathematics.

One of the basic problems of the book is the development of an analogue of the Poisson integral for the domains under consideration, and the solution of the Dirichlet problem for harmonic functions. A large part of the results presented in the book are due to the author himself. These results are established by sometimes very complicated direct computations, using an extensive algebraic machinery, and also the machinery of finite-dimensional representation theory which the author handles masterfully. Many of the auxiliary results, such as certain interesting algebraic identities, or the computation of integrals of functions of a matrix argument (e.g. the matrix analogue of the integral $\int_{-\infty}^{\infty}(1+x^2)^{-\alpha}dx$), undoubtedly have an independent interest.

In carrying out his investigations by direct computation the author unfortunately does not make use of the possibilities of the group-theoretic aspect of the problems. Yet this group-theoretic aspect would have made possible a clearer understanding of many of the results, and would sometimes have simplified their proofs. As an example, we indicate how the Poisson kernel can be defined in group-theoretic terms. Let $\mathfrak{R}$ be one of the domains considered in the book, and $\mathfrak{C}$ its characteristic manifold. Let z be a point in $\mathfrak{R}$ and C_z the group of those analytic automorphisms of $\mathfrak{R}$ which leave z invariant. It can be shown that the group C_z is transitive on $\mathfrak{C}$, i.e., transforms any point of $\mathfrak{C}$ into any other point. The measure on $\mathfrak{C}$ which is invariant under the transformations in C_z is then simply equal to the Poisson kernel.

The book of L. K. Hua is written rather concisely, and requires a good mathematical education and much attention from the reader. It is desirable that the reader be familiar with the basic notions of the theory of finite-dimensional group representations (e.g. the first three chapters of the book by F. Murnaghan, *Theory of group representations,* John Hopkins Press, 1938; Russian translation IL, 1950).

This monograph was originally published in China in a mimeographed edition shortly before the Third All-Soviet Mathematical Congress in Moscow (in 1956), at which the author took part. At the request of the Publishing House of Foreign Literature the author has specially revised and prepared the text of the book for publication in Russian.

At the end of the book there is a supplementary bibliography concerning questions related to the contents of this book.

M. I. Graev

FOREWORD

This monograph contains a number of the author's results concerning the theory of functions of several complex variables. Most of these results were published in Chinese in the journal Acta Mathematica Sinica, beginning with the year 1952. These results have been somewhat revised and supplemented by some later investigations of the author.

As a first sketch of this monograph one could consider the report I gave at the first general session of the Chinese Academy of Sciences, which was then partly repeated at the Third All-Soviet Mathematical Congress in Moscow, 1956.

I am very grateful to my colleagues Gun Šen,[1] Sjui I-pao, Čžun Tun-de and Lu-Ci-ken, especially the latter, who carefully checked through the manuscript and made a number of valuable comments.

I also wish to express my gratitude to the Publishing House of Foreign Literature, without whose cooperation this monograph could not have been published in Russian in such a short time.

I am especially grateful to colleagues M. A. Evgrafov and I. M. Graev for their hard work in the translation and editing of my book. The publication of this book is one of the concrete manifestations of the great friendship between the Soviet Union and the Chinese People's Republic.

L. K. Hua

[1] Editor's note. In the present English translation the Chinese names occurring in the Russian text are transliterated according to the rules for transliteration from Russian into English, i.e., as if they were Russian names.

INTRODUCTION

I. **Classical domains.** By a classical domain we shall understand an irreducible bounded symmetric domain (in the space of several complex variables) of one of the following four types:

(1) The domain $\mathfrak{R}_{\mathrm{I}}$ of $m \times n$ matrices with complex entries satisfying the condition

$$I^{(m)} - Z\overline{Z'} > 0.$$

Here $I^{(m)}$ is the identity matrix of order m, $\overline{Z'}$ is the complex conjugate of the transposed matrix Z'. ($H > 0$ for a hermitian matrix H denotes, as usual, that H is positive definite.)

(2) The domain $\mathfrak{R}_{\mathrm{II}}$ of symmetric matrices of order n (with complex entries) satisfying the condition

$$I^{(n)} - Z\overline{Z} > 0.$$

(3) The domain $\mathfrak{R}_{\mathrm{III}}$ of skew-symmetric matrices of order n (with complex entries) satisfying the condition

$$I^{(n)} + Z\overline{Z} > 0.$$

(4) The domain $\mathfrak{R}_{\mathrm{IV}}$ of n-dimensional ($n > 2$) vectors

$$z = (z_1, z_2, \cdots, z_n)$$

(z_k are complex numbers) satisfying the conditions[2]

$$|zz'|^2 + 1 - 2z\overline{z}' > 0, \qquad |zz'| < 1.$$

The complex dimension of these four domains is mn, $n(n+1)/2$, $n(n-1)/2$, n, respectively.

The author has shown (cf. L. K. Hua [3]) that $\mathfrak{R}_{\mathrm{IV}}$ can also be regarded as a homogeneous space of $2 \times n$ real matrices. Therefore, the study of all these domains can be reduced to a study of the geometry of matrices.

In 1935, E. Cartan [1] proved that there exist only six types of irreducible homogeneous bounded symmetric domains. Beside the above four types, there exist only two; their dimensions are 16 and 27. Of course

[2] Translator's note (n.b., unless otherwise noted, these words refer to the Russian translator). Here and throughout, the author considers a vector as a matrix of one row and n columns. So z' is a matrix of one column and n rows (the transpose of the matrix z).

these two types are rather special. The problem of the explicit description of these two types is still open.

The purpose of the present book is to study harmonic analysis on the classical domains. (The exact content of this harmonic analysis will be outlined later.)

II. **Characteristic manifolds.** Let $\mathfrak{R}$ be a bounded homogeneous domain in the $2n$-dimensional Euclidean space of n complex variables $z=(z_1, z_2, \cdots, z_n)$, and $f(z)$ an analytic function of z, regular in $\mathfrak{R}$. It is known that the maximum of the modulus of the function $f(z)$ is assumed on the boundary of $\mathfrak{R}$. Let $\mathfrak{C}$ be a manifold on the boundary of $\mathfrak{R}$ having the following properties:

(a) The modulus of every analytic function regular in $\mathfrak{R}$ assumes its maximum on $\mathfrak{C}$.

(b) For every point a on $\mathfrak{C}$ there exists a function $f(z)$, regular on $\mathfrak{R}$, such that the modulus of $f(z)$ assumes its maximum at $z=a$.

Such a manifold $\mathfrak{C}$ is called a characteristic manifold of the domain $\mathfrak{R}$. We should mention that $\mathfrak{C}$ is in general a proper subset of the boundary, and that the dimension of $\mathfrak{C}$ may be much less than $2n-1$. It is clear that $\mathfrak{C}$ is uniquely determined by $\mathfrak{R}$. It is easy to show that $\mathfrak{C}$ is closed, and that an analytic function which is regular in a neighborhood of each point of $\mathfrak{C}$ is uniquely determined by its values on $\mathfrak{C}$. Hence it follows that the real dimension of $\mathfrak{C}$ is not less than n. We shall denote by ξ the variable on $\mathfrak{C}$, and by $d\xi \overline{d\xi}'$ and $\dot{\xi}$ the metric and the element of volume of $\mathfrak{C}$.

Clearly, in the definition of $\mathfrak{C}$ it is enough to consider only linear functions instead of all analytic functions.

We describe the characteristic manifolds of the classical domains.

(1) $\mathfrak{C}_{\mathrm{I}}$ consists of the $m \times n$ matrices U satisfying the condition

$$U\overline{U}' = I^{(m)}.$$

(2) $\mathfrak{C}_{\mathrm{II}}$ consists of all symmetric unitary matrices of order n.

(3) $\mathfrak{C}_{\mathrm{III}}$ is defined differently for even and odd n. If n is even, then $\mathfrak{C}_{\mathrm{III}}$ consists of all skew-symmetric unitary matrices of order n. If n is odd, then $\mathfrak{C}_{\mathrm{III}}$ consists of all matrices of the form

$$UDU',$$

where U is an arbitrary unitary matrix and

$$D=\begin{pmatrix} 0 & 1 \\ -1 & 0 \end{pmatrix} \dot{+} \cdots \dot{+} \begin{pmatrix} 0 & 1 \\ -1 & 0 \end{pmatrix} \dot{+} 0.$$

(4) $\mathfrak{C}_{\mathrm{IV}}$ consists of the vectors $e^{i\theta}x$, where x is a real vector such that $xx'=1$.

The manifolds $\mathfrak{C}_{\mathrm{I}}$, $\mathfrak{C}_{\mathrm{II}}$, $\mathfrak{C}_{\mathrm{III}}$ and $\mathfrak{C}_{\mathrm{IV}}$ have real dimension $m(2n-m)$, $n(n+1)/2$, $n(n-1)/2+(1+(-1)^n)(n-1)/2$ and n, respectively.

These characteristic manifolds are homogeneous spaces. Furthermore, any point of $\mathfrak{C}$ can be carried into any other point of $\mathfrak{C}$ by a transformation leaving a given point of $\mathfrak{R}$ invariant. The general theory of harmonic analysis on homogeneous spaces has been developed earlier (cf. E. Cartan [1], H. Weyl [1]); however, the method presented in this book gives more precise and more useful results.

III. **Heuristic considerations.** Suppose that we have a sequence of analytic functions in $\mathfrak{R}$

$$\{\varphi_\nu(z)\}, \qquad \nu=0,1,2,\cdots,$$

such that any analytic function $f(z)$ in $\mathfrak{R}$ can be developed in a series

$$f(z)=\sum_{\nu=0}^{\infty} a_\nu \varphi_\nu(z),$$

convergent in $\mathfrak{R}$. We define the two infinite hermitian matrices

$$H_1=\left(\int_{\mathfrak{C}} \varphi_\nu(\xi)\overline{\varphi_\mu(\xi)}\,\dot{\xi}\right)_{\nu,\mu=0,1,2,\ldots}$$

and

$$H_2=\left(\int_{\mathfrak{R}} \varphi_\nu(z)\overline{\varphi_\mu(z)}\,\dot{z}\right)_{\nu,\mu=0,1,2,\ldots}$$

The basis $\{\varphi_\nu(z)\}$ can be chosen to be orthonormal, such that

$$\int_{\mathfrak{C}} \varphi_\nu(\xi)\overline{\varphi_\mu(\xi)}\,\dot{\xi}=\delta_{\nu\mu}$$

and

$$\int_{\mathfrak{R}} \varphi_\nu(z)\overline{\varphi_\mu(z)}\,\dot{z}=\lambda_\nu\delta_{\nu\mu}.$$

The eigenvalues $\lambda_0, \lambda_1, \lambda_2, \cdots$ are pseudoconformal invariants, i.e., they do not depend on the choice of the basis $\{\varphi_\nu(z)\}$ and are preserved under analytic mappings transforming $\mathfrak{R}$ and $\mathfrak{C}$ into $\mathfrak{R}_1$ and $\mathfrak{C}_1$, respectively.

The existence of a system $\{\varphi_\nu(z)\}$ is known from a theorem of H. Cartan

[1] on complete circular domains.[3]

Now setting

$$K(z, \overline{w}) = \sum_{\nu=0}^{\infty} \frac{\varphi_\nu(z)\overline{\varphi_\nu(w)}}{\lambda_\nu},$$

we obtain the Bergman kernel which has the following reproducing property. For any function $f(z)$ analytic in $\mathfrak{R}$ we have

$$f(z) = \int_{\mathfrak{R}} K(z, \overline{w}) f(w)\dot{w}.$$

Setting

$$H(z, \overline{\xi}) = \sum_{\nu=0}^{\infty} \varphi_\nu(z)\overline{\varphi_\nu(\xi)},$$

we obtain the Cauchy kernel of the domain $\mathfrak{R}$. This kernel has the reproducing property that for any analytic function $f(z)$ with a series development

$$f(\xi) = \sum_{\nu=0}^{\infty} a_\nu \varphi_\nu(\xi),$$

on $\mathfrak{C}$ we have

$$f(z) = \int_{\mathfrak{C}} H(z, \overline{\xi}) f(\xi)\dot{\xi}.$$

Setting

$$f(z) = u(z) H(z, \overline{w}),$$

we have

$$u(z) = \int_{\mathfrak{C}} \frac{H(z, \overline{\xi}) H(\xi, \overline{w})}{H(z, \overline{w})} u(\xi)\dot{\xi}.$$

The function

$$P(z, \xi) = \frac{H(z, \overline{\xi}) H(\xi, \overline{z})}{H(z, \overline{z})}$$

is called the Poisson kernel for the domain $\mathfrak{R}$. It is positive.

It is clear that the system of functions $\{\varphi_\nu(\xi)\}$, $\nu = 0, 1, 2, \cdots$ is not complete in the space of continuous functions on $\mathfrak{C}$. We complete it to a com-

---

[3] Translator's note. The domain $\mathfrak{R}$ in the space of several complex variables is said to be a circular domain (with center at the origin) if together with any point z in $\mathfrak{R}$ the point $ze^{i\varphi}$ is in $\mathfrak{R}$ for any real φ. If together with any point z in $\mathfrak{R}$ also the point $rze^{i\varphi}$ is in $\mathfrak{R}$ for any real φ and $0 \leq r \leq 1$, then $\mathfrak{R}$ is said to be a complete circular domain.

plete orthonormal system

$$\{\varphi_\nu(\xi)\}, \qquad \nu=0, \ \pm 1, \ \pm 2, \cdots,$$

and develop the function $P(z, \xi)$ into a Fourier series with respect to this new system

$$P(z, \xi) = \sum_{\nu=-\infty}^{\infty} \Phi_\nu(z) \overline{\varphi_\nu(\xi)}, \quad \Phi_\nu(z) = \int_{\mathfrak{C}} P(z, \xi) \varphi_\nu(\xi) \dot{\xi}.$$

If

$$\lim_{z \to \xi} \Phi_\nu(z) = \varphi_\nu(\xi),$$

then the functions on $\mathfrak{C}$ having a Fourier series development

$$\varphi(\xi) = \sum_{\nu=-\infty}^{\infty} c_\nu \varphi_\nu(\xi), \quad c_\nu = \int_{\mathfrak{C}} \varphi(\xi) \overline{\varphi_\nu(\xi)} \dot{\xi}$$

can be put in correspondence with the class of functions

$$\Phi(z) = \int_{\mathfrak{C}} P(z, \xi) \varphi(\xi) \dot{\xi} = \sum_{\nu=-\infty}^{\infty} c_\nu \Phi_\nu(z),$$

which we shall call harmonic functions in the domain $\mathfrak{R}$.

The harmonic functions can also be defined as solutions of a second order partial differential equation. This equation can be obtained from the following considerations.

The Bergman kernel yields a Riemannian metric on the space $\mathfrak{R}$:

$$d\bar{d} \ln K(z, \bar{z}) = \sum_{i=1}^{n} \sum_{j=1}^{n} \frac{\partial^2}{\partial z_i \partial \bar{z}_j} \ln K(z, \bar{z}) \, dz_i \, d\bar{z}_j = \sum_{i, j=1}^{n} h_{ij} \, dz_i \, d\bar{z}_j.$$

Corresponding to the tensor h^{ij} we have a differential operator

$$\sum_{i=1}^{n} \sum_{j=1}^{n} h^{ij} \frac{\partial^2}{\partial z_i \partial \bar{z}_j},$$

which we call the Laplace operator of this space.

We can talk about a Dirichlet problem with respect to this operator.

All the facts mentioned in this section will be treated in detail later in the book.

IV. **Remarks on the methods to be used.**

(a) *The machinery of group representation theory.* It is known that the classical domains of all four types are complete circular domains (cf. the remark on p. 7). For complete circular domains it can be assumed that the group of motions leaving the origin fixed consists of linear transformations of the form

$$w = zU,$$

where U is a unitary matrix. We denote this group by Γ_0. The system of all monomials $z_1^{l_1} z_2^{l_2} \cdots z_n^{l_n}$ is a complete system in the space of functions analytic on $\mathfrak{R}$, and the monomials of different degrees (i.e., $l_1 + \cdots + l_n = l_1' + \cdots + l_n'$) are orthogonal to one another. Therefore, our basic problem is to split the set of homogeneous monomials of the same degree into an orthonormal system. More exactly, let $z = (z_1, z_2, \cdots, z_n)$, and let $z^{[l]}$ be the vector with components

$$\sqrt{\frac{l!}{l_1!\, l_2! \, \ldots \, l_n!}}\, z_1^{l_1} z_2^{l_2} \ldots z_n^{l_n}, \qquad l = l_1 + l_2 + \ldots + l_n,$$

in $n(n+1)\cdots(n+l-1)/l!$-dimensional space. The transformation $w = zU$ induces a transformation

$$w^{[l]} = z^{[l]}\, U^{[l]},$$

where $U^{[l]}$ is the lth symmetric Kronecker power[4] of U. In order to find the orthogonal components of $z^{[l]}$-space, we have to solve the problem of decomposing the representation $U^{[l]}$ into irreducible components. This corresponds to the following problem in the representation theory of linear groups.

We start with the group $GL(n)$ of all nondegenerate matrices of order n. Let $f_1, \cdots, f_n$ be integers such that $f_1 \geqq f_2 \geqq \cdots \geqq f_n \geqq 0$. To every element X in $GL(n)$ there corresponds a matrix

$$A_{f_1, \cdots, f_n}(X)$$

in the representation with signature $(f_1, f_2, \cdots, f_n)$ of $GL(n)$.[5] The degree of this representation will be denoted by $N = N(f_1, f_2, \cdots, f_n)$. For unitary X the matrix $A_{f_1, \cdots, f_n}(X)$ is unitary. Let

$$B_{g_1, \cdots, g_N}(Y)$$

be a representation of the group $GL(N)$. Clearly,

$$B_{g_1, \cdots, g_N}(A_{f_1, \cdots, f_n}(X))$$

is again a representation of the group $GL(n)$. The problem is to find the irreducible components of this composite representation. This problem is

[4] Translator's note. Cf. F. Murnaghan [**1**, p. 104].
[5] Translator's note. Cf. H. Weyl [**2**, p. 181].

in general very complicated, but in the two special cases which we shall need, it can be solved completely.[6]

(b) *Polar coordinates for matrices.* After obtaining a complete orthogonal system using the methods of representation theory, we shall encounter another difficulty; that of computing the normalizing constants. (This is easier for $\mathfrak{C}$ than $\mathfrak{R}$.) In order to overcome this difficulty, we shall introduce polar coordinates for matrices. We explain what this means for the example of symmetric matrices.

Every complex symmetric matrix Z of order n can be represented in the form

$$Z = U\Lambda U',$$

where U is unitary and Λ is a real diagonal matrix $\Lambda = [\lambda_1, \lambda_2, \cdots, \lambda_n]$ with $\lambda_1 \geqq \lambda_2 \geqq \cdots \geqq \lambda_n \geqq 0$ (cf. L. K. Hua [1]). Now let $\{U\}$ be the set of cosets of the unitary group with respect to its subgroup consisting of the 2^n diagonal matrices

$$[\pm 1, \pm 1, \cdots, \pm 1].$$

Then to every symmetric matrix there corresponds an element from U and a diagonal matrix Λ. This correspondence is one-to-one for almost all matrices (i.e., with the possible exception of a manifold of lower dimension). The matrix Λ is the "modulus" of the matrix Z, and U is its "argument". In this book we shall compute the Jacobian of the transformation from "Cartesian coordinates Z" to the "polar coordinates $(\{U\}, \Lambda)$". We shall also obtain some analogous results about "polar coordinates" for other types of matrices, although some of these are not directly connected with the fundamental problems of this book.

(c) *Computation of certain integrals.* In this book there are, of course, many interesting integrals; here we mention only one of them.

For $\alpha > 1$, $\alpha + \beta > -n$ we have

$$\int_{\mathfrak{R}_{IV}} \left(1 - \bar{z}z' - \sqrt{(\bar{z}z')^2 - |zz'|^2}\right)^{\alpha} \left(1 - \bar{z}z' + \sqrt{(\bar{z}z')^2 - |zz'|^2}\right)^{\beta} \dot{z}$$

$$= \frac{\pi^n \Gamma(\alpha + 1)}{2^{n-1}(\alpha + \beta + n)\Gamma(\alpha + n)}.$$

---

[6] The author is grateful to Professor H. Duan for the information that the problem was also solved in these cases by R. M. Thrall [1]. His method is quite different from the one presented in this book.

Consequently, for $\lambda<1$,

$$\int_{\mathfrak{R}_{IV}} (1-2\bar{z}z'+|zz'|^2)^{-\lambda}\dot{z}=\frac{\pi^n\Gamma(1-\lambda)}{2^{n-1}(n-2\lambda)\Gamma(n-\lambda)}.$$

This formula, incidentally, solves a problem posed by the author in 1946 (cf. L. K. Hua [**3**]): In $\mathfrak{R}_{IV}$ the index of convergence for Poincaré series is not less than $(n-1)/n$, and this estimate cannot be improved.

V. **Applications to Representation Theory.** In the preceding section we touched on the question as to how the machinery of representation theory can be used in the theory of functions of several complex variables. Here we show, conversely, how the results of our investigations can be applied in representation theory.

For greater clarity we shall first talk about harmonic functions in $\mathfrak{R}_I$ for $m=n$. The Laplace equation in $\mathfrak{R}_I$ has the form

$$\sum_{\alpha,\beta=1}^{n}\sum_{j,k=1}^{n}\left(\delta_{\alpha\beta}-\sum_{l=1}^{n} z_{l\alpha}\bar{z}_{l\beta}\right)\left(\delta_{jk}-\sum_{\gamma=1}^{n} z_{j\gamma}\bar{z}_{k\gamma}\right)\frac{\partial^2 u}{\partial z_{j\alpha}\,\partial\bar{z}_{k\beta}}=0.$$

The functions $u(Z)$ satisfying this equation on the closure of $\mathfrak{R}_I$ will be called harmonic in $\mathfrak{R}_I$. Those among them which have continuous boundary values on $\mathfrak{C}_I$ form a class which we shall denote by $\mathfrak{H}$. The solution of the Dirichlet problem on $\mathfrak{R}_I$ gives the following result.

If we are given a continuous function $\varphi(U)$ on the unitary group $\mathfrak{C}_I$, then there exists one and only one harmonic function $u(Z)$ satisfying the condition

$$\lim_{Z\to U} u(Z)=\varphi(U).$$

This function can be found from the Poisson formula

$$u(Z)=\frac{1}{V(\mathfrak{C})}\int_{\mathfrak{C}}\frac{\{\det(I-Z\bar{Z}')\}^n}{|\det(I-Z\bar{U}')|^{2n}}\varphi(U)\dot{U}.$$

A series decomposition of the poisson kernel

$$\frac{1}{V(\mathfrak{C})}\frac{\{\det(I-Z\bar{Z}')\}^n}{|\det(I-Z\bar{U}')|^{2n}}$$

can be obtained as follows. Let

$$A_f(U)=A_{f_1,\cdots,f_n}(U)=(a^f_{ij}(U)),\quad 1\leqq i,\ j\leqq N(f_1,\cdots,f_n)=N(f).$$

The sequence

$$\varphi^f_{ij}(U)=\sqrt{\frac{N(f)}{V(\mathfrak{C})}}\,a^f_{ij}(U),\quad i,\ j=1,\ 2,\ \ldots,\ N(f),$$
$$-\infty<f_n\leqslant\cdots\leqslant f_2\leqslant f_1<\infty,$$

is orthonormal on the unitary group $\mathfrak{C}_{\mathrm{I}}$, and the desired series is of the form

$$\sum_f\sum_{i,j}\Phi^f_{ij}(Z)\,\overline{\varphi^f_{ij}(U)},\qquad \Phi^f_{ij}(Z)=\frac{1}{V(\mathfrak{C})}\int_{\mathfrak{C}}\frac{\{\det(I-Z\bar{Z}')\}^n}{|\det(I-Z\bar{U}')|^{2n}}\varphi^f_{ij}(U)\dot{U},$$

where

$$\lim_{Z\to U}\Phi^f_{ij}(Z)=\varphi^f_{ij}(U).$$

For every continuous function $\varphi(U)$ we can set up the formal Fourier series

$$\sum_f\sum_{i,j}c^f_{ij}\varphi^f_{ij}(U),\qquad c^f_{ij}=\frac{1}{V(\mathfrak{C})}\int_{\mathfrak{C}}\varphi(U)\,\overline{\varphi^f_{ij}(U)}\,\dot{U},$$

which in general may not converge to $\varphi(U)$. Still we have

$$\varphi(U)=\lim_{Z\to U}\sum_f\sum_{i,j}c^f_{ij}\Phi^f_{ij}(Z).$$

This means that the formal Fourier series is Abel summable to $\varphi(U)$. Thus we obtain the theorem that the Fourier series of any continuous function on the unitary group is Abel summable.

As a consequence we have the Peter-Weyl approximation theory for an arbitrary finite-dimensional compact group and the approximation theorem for homogeneous spaces. It is clear that these results are more precise than in the previously known formulations.

We make the following additional remarks.

The harmonic functions of class $\mathfrak{H}$ also satisfy the system of n^2 differential equations

$$\sum_{\alpha,\beta=1}^{n}\left(\delta_{\alpha\beta}-\sum_{h=1}^{n}\bar{z}_{h\alpha}z_{h\beta}\right)\frac{\partial^2 u}{\partial\bar{z}_{i\alpha}\,\partial z_{j\beta}}=0,\qquad 1\leqslant i,j\leqslant n.$$

This appears to be very striking and suggests the possibility in general of splitting a partial differential equation into a system of partial differential equations.

Furthermore, the domain of convergence of the series

$$\sum_{f} \sum_{i, j} c_{ij}^{f} \Phi_{ij}^{f}(Z)$$

does not necessarily coincide with $\mathfrak{R}_{\mathrm{I}}$. But if it converges outside $\mathfrak{R}_{\mathrm{I}}$ then the Laplace equation cannot be elliptic there, and the sum of the series must be the solution of an equation of mixed type.

Chapter I

ALGEBRAIC MACHINERY

1.1. **Algebraic identities.** Everywhere in what follows we shall use the notation

$$D(x_1, \ldots, x_n) = \prod_{1 \leqslant i < j \leqslant n} (x_i - x_j), \qquad n \geqslant 2.$$

THEOREM 1.1.1. *The identity*

$$\sum_{i_1, \ldots, i_n} \delta^{1\,2\,\ldots\,n}_{i_1 i_2 \ldots i_n} \frac{x_{i_1}^{n-1} x_{i_2}^{n-2} \cdots x_{i_{n-1}}}{\left(1 - x_{i_1}^2\right)\left(1 - x_{i_1}^2 x_{i_2}^2\right) \cdots \left(1 - x_{i_1}^2 \cdots x_{i_n}^2\right)}$$
$$= D(x_1, \ldots, x_n) \prod_{1 \leqslant i \leqslant j \leqslant n} (1 - x_i x_j)^{-1}. \tag{1.1.1}$$

holds. Here $i_1, i_2, \cdots, i_n$ are all possible permutations of the numbers $1, 2, \cdots, n$ and $\delta^{12\cdots n}_{i_1 i_2 \cdots i_n}$ is equal to 1 or -1 according as the permutation is even or odd.

THEOREM 1.1.2. *Let $\nu = [n/2]$. We have the identity*

$$\sum_{i_1, \ldots, i_n} \delta^{12\,\ldots\,n}_{i_1 i_2 \ldots i_n} \frac{x_{i_1}^{n-1} x_{i_2}^{n-2} \cdots x_{i_{n-1}}}{\left(1 - x_{i_1} x_{i_2}\right)\left(1 - x_{i_1} x_{i_2} x_{i_3} x_{i_4}\right) \cdots \left(1 - x_{i_1} \cdots x_{i_{2\nu}}\right)}$$
$$= D(x_1, \ldots, x_n) \prod_{1 \leqslant i < j \leqslant n} (1 - x_i x_j)^{-1} \tag{1.1.2}$$

For the proof of these two identities we need two simple results on Vandermonde determinants. It is well known that

$$D(x_1, \ldots, x_n) = (-1)^{\frac{n(n-1)}{2}} \det \left| x_i^{j-1} \right|_1^n. \tag{1.1.3}$$

Let

$$D_i = D(x_1, \cdots, x_{i-1}, 0, x_{i+1}, \cdots, x_n).$$

Then we have

$$\sum_{i=1}^{n} x_i^l D_i = \begin{cases} D(x_1, \ldots, x_n) & \text{for} \quad l=0, \\ 0 & \text{for} \quad 1 \leqslant l \leqslant n-1, \\ (-1)^{n-1} D(x_1, \ldots, x_n)\, x_1 \ldots x_n & \text{for} \quad l=n. \end{cases} \tag{1.1.4}$$

Then since

$$\begin{vmatrix} \frac{x_1}{1-x_1} & \frac{x_2}{1-x_2} & \cdots & \frac{x_n}{1-x_n} \\ x_1 & x_2 & \cdots & x_n \\ \cdots & \cdots & \cdots & \cdots \\ x_1^{n-1} & x_2^{n-1} & \cdots & x_n^{n-1} \end{vmatrix}$$

$$= \frac{1}{\prod_{i=1}^{n}(1-x_i)} \begin{vmatrix} x_1 & x_2 & \cdots & x_n \\ x_1(1-x_1) & x_2(1-x_2) & \cdots & x_n(1-x_n) \\ \cdots & \cdots & \cdots & \cdots \\ x_1^{n-1}(1-x_1) & x_2^{n-1}(1-x_2) & \cdots & x_n^{n-1}(1-x_n) \end{vmatrix}$$

$$= (-1)^{n-1+\frac{n(n-1)}{2}} \frac{x_1 \ldots x_n}{\prod_{i=1}^{n}(1-x_i)} D(x_1, \ldots, x_n),$$

we have

$$\sum_{i=1}^{n} D_i \frac{x_i}{1-x_i} = (-1)^{n-1} \frac{x_1 \ldots x_n}{\prod_{i=1}^{n}(1-x_i)} D(x_1, \ldots, x_n). \tag{1.1.5}$$

Changing x_i to $-x_i$ in this formula we obtain

$$\sum_{i=1}^{n} D_i \frac{x_i}{1+x_i} = \frac{x_1 \ldots x_n}{\prod_{i=1}^{n}(1+x_i)} D(x_1, \ldots, x_n). \tag{1.1.6}$$

Setting $l=0$ in (1.1.4) and adding (1.1.5) we find

$$\sum_{i=1}^{n} \frac{D_i}{1-x_i} = D(x_1, \ldots, x_n)\left\{1 + (-1)^{n-1} \frac{x_1 \ldots x_n}{\prod_{i=1}^{n}(1-x_i)}\right\}, \tag{1.1.7}$$

whence it follows without difficulty that

$$\sum_{i=1}^{n} \frac{D_i}{1+x_i} = D(x_1, \ldots, x_n)\left\{1 - \frac{x_1 \ldots x_n}{\prod_{i=1}^{n}(1+x_i)}\right\}. \tag{1.1.8}$$

On the basis of these results we now prove Theorems 1.1.1 and 1.1.2.

PROOF OF 1.1.1. For $n=2$ the left-hand side of 1.1.1 is equal to

$$\frac{x_1}{(1-x_1^2)(1-x_1^2x_2^2)} - \frac{x_2}{(1-x_2^2)(1-x_1^2x_2^2)} = \frac{(x_1-x_2)(1+x_1x_2)}{(1-x_1^2)(1-x_2^2)(1-x_1^2x_2^2)}$$

$$= \frac{x_1-x_2}{(1-x_1^2)(1-x_1x_2)(1-x_2^2)} = D(x_1, x_2) \prod_{1 \leqslant i \leqslant j \leqslant 2} (1-x_ix_j)^{-1},$$

so, for $n=2$ the theorem is true. We proceed by induction. Assume the theorem is true for $n-1$. Then

$$\sum_{i_1, \ldots, i_n} \delta^{1 \ldots n}_{i_1 \ldots i_n} \frac{x_{i_1}^{n-1} x_{i_2}^{n-2} \ldots x_{i_{n-1}}}{(1-x_{i_1}^2)(1-x_{i_1}^2x_{i_2}^2) \ldots (1-x_{i_1}^2 \ldots x_{i_n}^2)}$$

$$= \sum_{a=1}^{n} (-1)^{n-a} \sum_{i_1, \ldots, i_{n-1}} \delta^{1 \ldots a-1, a+1, \ldots n}_{i_1 \ldots i_{n-1}}$$

$$\times \frac{x_{i_1}^{n-1} x_{i_2}^{n-2} \ldots x_{i_{n-1}}}{(1-x_{i_1}^2) \ldots (1-x_{i_1}^2 \ldots x_{i_n}^2)} = \sum_{a=1}^{n} (-1)^{n-a} \frac{x_1x_2 \ldots x_n}{x_a(1-x_1^2x_2^2 \ldots x_n^2)}$$

$$\times \sum_{i_1, \ldots, i_{n-1}} \delta^{1 \ldots a-1, a+1 \ldots n}_{i_1 \ldots i_{n-1}} \frac{x_{i_1}^{n-2} x_{i_2}^{n-3} \ldots x_{i_{n-2}}}{(1-x_{i_1}^2) \ldots (1-x_{i_1}^2 \ldots x_{i_{n-1}}^2)}$$

$$= \sum_{a=1}^{n} (-1)^{n-a} \frac{x_1 \ldots x_n}{x_a(1-x_1^2 \ldots x_n^2)} \cdot \frac{D(x_1, \ldots, x_{a-1}, x_{a+1}, \ldots, x_n)}{\prod_{\substack{1 \leqslant i \leqslant j \leqslant n \\ i \neq a,\, j \neq a}} (1-x_ix_j)}$$

$$= \frac{1}{(1-x_1^2 \ldots x_n^2) \prod_{1 \leqslant i \leqslant j \leqslant n} (1-x_ix_j)} \sum_{a=1}^{n} D_a \prod_{i=1}^{n} (1-x_ix_a). \tag{1.1.9}$$

Since

$$\prod_{i=1}^{n} (1-x_ix_a) = \sum_{l=0}^{n} \sigma_l x_a^l,$$

where σ_l denotes the lth elementary symmetric function of $x_1, \cdots, x_n$, by (1.1.4) we have

$$\sum_{a=1}^{n} D_a \prod_{i=1}^{n} (1-x_i x_a) = \sum_{a=1}^{n} D_a \sum_{l=0}^{n} \sigma_l x_a^l = \sum_{l=0}^{n} \sigma_l \sum_{a=1}^{n} D_a x_a^l$$
$$= D(x_1,\ldots,x_n) + (-1)^n x_1 \ldots x_n \cdot (-1)^{n-1} D(x_1,\ldots,x_n)\, x_1 \ldots x_n$$
$$= D(x_1, \ldots, x_n)\left(1 - x_1^2 \ldots x_n^2\right).$$

Substituting into (1.1.9) we get (1.1.1).

PROOF OF 1.1.2. For $n=2$ the identity (1.1.2) is obvious. We now perform an induction. If the theorem is true for $n-1$, then

$$\sum_{i_1, \ldots, i_n} \delta_{i_1 i_2 \ldots i_n}^{1\,2\,\ldots\,n} \frac{x_{i_1}^{n-1} x_{i_2}^{n-2} \ldots x_{i_{n-1}}}{\left(1 - x_{i_1} x_{i_2}\right)\left(1 - x_{i_1} x_{i_2} x_{i_3} x_{i_4}\right) \cdots \left(1 - x_{i_1} x_{i_2} \cdots x_{i_{2\nu}}\right)}$$
$$= \sum_{a=1}^{n} (-1)^{n-a} \sum_{i_1, \ldots, i_{n-1}} \delta_{i_1 \ldots i_{n-1}}^{1 \ldots a-1, a+1 \ldots n}$$
$$\times \frac{x_{i_1}^{n-1} x_{i_2}^{n-2} \ldots x_{i_{n-1}}}{\left(1 - x_{i_1} x_{i_2}\right) \cdots \left(1 - x_{i_1} \cdots x_{i_{2\nu}}\right)} = \sum_{a=1}^{n} (-1)^{n-a} \frac{x_1 \ldots x_n}{x_a (1 - \varepsilon x_1 \ldots x_n)}$$
$$\times \sum_{i_1, \ldots, i_{n-1}} \delta_{i_1 \ldots i_{n-1}}^{1 \ldots a-1, a+1 \ldots n} \frac{x_{i_1}^{n-2} x_{i_2}^{n-3} \ldots x_{i_{n-2}}}{\left(1 - x_{i_1} x_{i_2}\right)\left(1 - x_{i_1} x_{i_2} x_{i_3} x_{i_4}\right) \cdots}, \tag{1.1.10}$$

where ϵ is equal to 0 or 1 according to whether n is even or odd. By the induction hypothesis, the inner sum is equal to

$$\frac{D(x_1, \ldots, x_{a-1}, x_{a+1}, \ldots, x_n)}{\prod_{1 \leqslant i < j \leqslant n} (1 - x_i x_j)} \cdot \frac{\prod_{i=1}^{n} (1 - x_i x_a)}{1 - x_a^2}.$$

Substituting into (1.1.10) it follows that the left-hand side of (1.1.2) is equal to

$$\sum_{a=1}^{n} (-1)^{n-a} \frac{x_1 \ldots x_n D(x_1, \ldots, x_{a-1}, x_{a+1}, \ldots, x_n)}{x_a (1 - \varepsilon x_1 \ldots x_n) \prod_{1 \leqslant i < j \leqslant n} (1 - x_i x_j)} \cdot \frac{\prod_{i=1}^{n} (1 - x_i x_a)}{1 - x_a^2}$$
$$= (1 - \varepsilon x_1 \ldots x_n)^{-1} \prod_{1 \leqslant i < j \leqslant n} (1 - x_i x_j)^{-1} \sum_{a=1}^{n} \frac{D_a}{1 - x_a^2} \prod_{i=1}^{n} (1 - x_i x_a)$$
$$= (1 - \varepsilon x_1 \ldots x_n)^{-1} \prod_{1 \leqslant i < j \leqslant n} (1 - x_i x_j)^{-1} \sum_{l=0}^{n} \sigma_l \sum_{a=1}^{n} \frac{D_a x_a^l}{1 - x_a^2}. \tag{1.1.11}$$

In considering the inner sum in (1.1.11) we distinguish the following three cases:

(1) For odd l, by

$$\frac{x_a - x_a^l}{1-x_a^2} = x_a(1+x_a^2+\ \dots\ +x_a^{l-3}),$$

we have

$$-\frac{x_a^l}{1-x_a^2} = x_a(1+x_a^2+\ \dots\ +x_a^{l-3}) - \frac{1}{2}\left(\frac{x_a}{1-x_a}+\frac{x_a}{1+x_a}\right).$$

From (1.1.4), (1.1.5) and (1.1.6) we get at once

$$\sum_{a=1}^{n}\frac{D_a x_a^l}{1-x_a^2} = \frac{1}{2}\sum_{a=1}^{n}\left(\frac{x_a}{1+x_a}+\frac{x_a}{1-x_a}\right)D_a$$
$$= \frac{1}{2}D(x_1, \dots, x_n)\, x_1 \dots x_n\left\{(-1)^{n-1}\prod_{i=1}^{n}(1-x_i)^{-1}+\prod_{i=1}^{n}(1+x_i)^{-1}\right\}. \tag{1.1.12}$$

(2) For positive even l, since

$$\frac{1-x_a^l}{1-x_a^2} = 1+x_a^2+\ \dots\ +x_a^{l-2},$$

we have

$$-\frac{x_a^l}{1-x_a^2} = 1+x_a^2+\ \dots\ +x_a^{l-2} - \frac{1}{2}\left(\frac{1}{1-x_a}+\frac{1}{1+x_a}\right).$$

From (1.1.4), (1.1.7) and (1.1.8) we obtain

$$\sum_{a=1}^{n}\frac{D_a x_a^l}{1-x_a^2} = -\sum_{a=1}^{n}D_a + \frac{1}{2}\sum_{a=1}^{n}\left(\frac{1}{1-x_a}+\frac{1}{1+x_a}\right)D_a$$
$$= -D(x_1, \dots, x_n) + \frac{1}{2}D(x_1, \dots, x_n)$$
$$\times\left\{1+(-1)^{n-1}x_1 \dots x_n\prod_{i=1}^{n}(1-x_i)^{-1}\right.$$
$$\left.+1-x_1 \dots x_n\prod_{i=1}^{n}(1+x_i)^{-1}\right\} = \frac{1}{2}D(x_1, \dots, x_n)$$
$$\times x_1 \dots x_n\left\{(-1)^{n-1}\prod_{i=1}^{n}(1-x_i)^{-1}-\prod_{i=1}^{n}(1+x_i)^{-1}\right\}. \tag{1.1.13}$$

(3) For $l=0$ we have

$$\sum_{a=1}^{n}\frac{D_a}{1-x_a^2}=\frac{1}{2}\sum_{a=1}^{n}\Big(\frac{1}{1-x_a}+\frac{1}{1+x_a}\Big)D_a=D(x_1,\ \ldots,\ x_n)$$

$$+\frac{1}{2}D(x_1,\ldots,x_n)\,x_1\ldots x_n\Big\{(-1)^{n-1}\prod_{i=1}^{n}(1-x_i)^{-1}-\prod_{i=1}^{n}(1+x_i)^{-1}\Big\}. \tag{1.1.14}$$

From (1.1.12), (1.1.13) and (1.1.14) we obtain

$$\sum_{l=0}^{n}\sigma_l\sum_{a=1}^{n}\frac{D_a x_a^l}{1-x_a^2}=D(x_1,\ldots,x_n)$$

$$+\frac{1}{2}D(x_1,\ldots,\hat{x}_n)\,x_1\ldots x_n\Big\{(-1)^{n-1}\prod_{i=1}^{n}(1-x_i)^{-1}-\prod_{i=1}^{n}(1+x_i)^{-1}\Big\}$$

$$+\frac{1}{2}D(x_1,\ldots,x_n)\,x_1\ldots x_n\Big\{(-1)^{n-1}\prod_{i=1}^{n}(1-x_i)^{-1}-\prod_{i=1}^{n}(1+x_i)^{-1}\Big\}$$

$$\times\sum_{\substack{l=1\\(l=2k)}}^{n}\sigma_l+\frac{1}{2}D(x_1,\ \ldots,\ x_n)\,x_1\ \ldots\ x_n\Big\{(-1)^{n-1}\prod_{i=1}^{n}(1-x_i)^{-1}$$

$$+\prod_{i=1}^{n}(1+x_i)^{-1}\Big\}\sum_{\substack{l=1\\(l=2k+1)}}^{n}\sigma_l=D(x_1,\ldots,x_n)+\frac{1}{2}D(x_1,\ldots,x_n)\,x_1\ldots x_n$$

$$\times\Big\{(-1)^{n-1}\prod_{i=1}^{n}(1-x_i)^{-1}\sum_{l=0}^{n}\sigma_l+\prod_{i=1}^{n}(1+x_i)^{-1}\sum_{l=1}^{n}(-1)^{l+1}\sigma_l\Big\}$$

$$=D(x_1,\ \ldots,\ x_n)\Big\{1+\frac{-1+(-1)^{n-1}}{2}\,x_1\ \ldots\ x_n\Big\}. \tag{1.1.15}$$

Substituting (1.1.15) into (1.1.11) and noting that $(-1)^{n-1}-1=-2\epsilon$, we find

$$(1-\varepsilon x_1\ \ldots\ x_n)^{-1}\prod_{1\leqslant i<j\leqslant n}(1-x_ix_j)^{-1}D(x_1,\ \ldots,\ x_n)(1-\varepsilon x_1\ \ldots\ x_n)$$

$$=D(x_1,\ \ldots,\ x_n)\prod_{1\leqslant i<j\leqslant n}(1-x_ix_j)^{-1}, \tag{1.1.16}$$

which proves Theorem 1.1.2.

We also prove the following relatively simple identity.

THEOREM 1.1.3.

$$\det\left|\frac{1}{x_i+y_j}\right|_1^n = D(x_1, \ldots, x_n)\,D(y_1, \ldots, y_n)\prod_{i=1}^{n}\prod_{j=1}^{n}(x_i+y_j)^{-1}. \tag{1.1.17}$$

PROOF. We subtract the first row from the second, third, $\cdots$, nth row and use the identity

$$\frac{1}{x_l+y_k}-\frac{1}{x_1+y_k}=\frac{x_1-x_l}{(x_1+y_k)(x_l+y_k)};\quad l=2, \ldots, n,\ k=1, \ldots, n.$$

After these transformations the determinant is of the form

$$\frac{(x_1-x_2)\ldots(x_1-x_n)}{\prod_{k=1}^{n}(x_1+y_k)}\begin{vmatrix} 1 & 1 & \cdots & 1 \\ \frac{1}{x_2+y_1} & \frac{1}{x_2+y_2} & \cdots & \frac{1}{x_2+y_n} \\ \cdot & \cdot & \cdots & \cdot \\ \frac{1}{x_n+y_1} & \frac{1}{x_n+y_2} & \cdots & \frac{1}{x_n+y_n} \end{vmatrix}. \tag{1.1.18}$$

We subtract the first column of this determinant from its second, third, $\cdots$, nth column. After this the determinant in (1.1.18) becomes

$$(y_1-y_2)\ldots(y_1-y_n)\prod_{j=2}^{n}(x_j+y_1)^{-1}\det\left|\frac{1}{x_i+y_j}\right|_2^n.$$

Now the theorem can at once be proved by induction.

Changing $x_1, x_2, \cdots, x_n$ to $-x_1, -x_2, \cdots, -x_n$ after some simple transformation we get the following result.

THEOREM 1.1.4.

$$\det\left|(1-x_iy_j)^{-1}\right|_1^n = D(x_1, \ldots, x_n)\,D(y_1, \ldots, y_n)\prod_{i=1}^{n}\prod_{j=1}^{n}(1-x_iy_j)^{-1}.$$ [7]

[7] Translator's note. This identity is known as Cauchy's lemma. For another proof of it cf. Weyl [2, p. 276].

1.2. **Power series identities.**

THEOREM 1.2.1. *Let the power series*

$$f_i(z)=\sum_{k=0}^{\infty} a_k^{(i)} z^k \tag{1.2.1}$$

converge for $|z|<\rho$. *Then for* $|z_1|<\rho, \cdots, |z_n|<\rho$ *the following identity holds*

$$\det|f_i(z_j)|_1^n=\sum_{l_1>l_2>\cdots>l_n\geqslant 0} \det\left|a_{l_j}^{(i)}\right|_{i,j=1}^n \det\left|z_i^{l_j}\right|_{i,j=1}^n. \tag{1.2.2}$$

PROOF. By the definition of a determinant the left-hand side of (1.2.2) equals

$$\sum_{i_1\ldots i_n} \delta_{i_1\ldots i_n}^{1\ldots n} f_{i_1}(z_1)\cdots f_{i_n}(z_n)$$

$$=\sum_{i_1\ldots i_n} \delta_{i_1\ldots i_n}^{1\ldots n} \sum_{k_1=0}^{\infty}\cdots\sum_{k_n=0}^{\infty} a_{k_1}^{(i_1)}\cdots a_{k_n}^{(i_n)} z_1^{k_1}\cdots z_n^{k_n}$$

$$=\sum_{k_1=0}^{\infty}\cdots\sum_{k_n=0}^{\infty}\left(\sum_{i_1\ldots i_n} \delta_{i_1\ldots i_n}^{1\ldots n} a_{k_1}^{(i_1)}\cdots a_{k_n}^{(i_n)}\right) z_1^{k_1}\cdots z_n^{k_n}$$

$$=\sum_{k_1=0}^{\infty}\cdots\sum_{k_n=0}^{\infty} z_1^{k_1}\cdots z_n^{k_n}\det\left|a_{k_j}^{(i)}\right|_{i,j=1}^n$$

$$=\sum_{l_1>l_2>\cdots>l_n\geqslant 0}\det\left|a_{l_j}^{(i)}\right|_{i,j=1}^n \sum_{i_1\ldots i_n} \delta_{i_1\ldots i_n}^{1\ldots n} z_1^{l_{i_1}}\cdots z_n^{l_{i_n}}$$

$$=\sum_{l_1>\cdots>l_n\geqslant 0}\det\left|a_{l_j}^{(i)}\right|_{i,j=1}^n \det\left|z_i^{l_j}\right|_{i,j=1}^n,$$

which was to be proved.

Setting $f_i(z)=f(x_i z)$ we get the following special case.

THEOREM 1.2.2. *Let the power series*

$$f(z)=a_0+a_1z+a_2z^2+\cdots \tag{1.2.3}$$

converge for $|z|<\rho$. *Then for* $|x_i y_j|<\rho$, $i, j=1,\cdots,n$, *we have*

$$\det|f(x_iy_j)|_1^n = \sum_{l_1>\cdots>l_n\geqslant 0} a_{l_1}a_{l_2}\ldots a_{l_n}\det|x_i^{l_j}|_{i,j=1}^n \det|y_i^{l_j}|_{i,j=1}^n. \quad (1.2.4)$$

In particular, for $f(z)=(1-z)^{-1}$ we have the following formula

$$\sum_{l_1>\cdots>l_n\geqslant 0}\det|x_i^{l_j}|_{i,j=1}^n \det|y_i^{l_j}|_{i,j=1}^n = \det|(1-x_iy_j)^{-1}|_1^n, \quad (1.2.5)$$

and by Theorem 1.1.4 the latter determinant equals

$$D(x_1,\ldots,x_n)D(y_1,\ldots,y_n)\prod_{i=1}^n\prod_{j=1}^n(1-x_iy_j)^{-1}.$$

Now we set $x_n=0$ in (1.2.5). Then all terms with $l_n>0$ vanish, and we get

$$\sum_{l_1>\cdots>l_{n-1}>0}\det|x_i^{l_j}|_{i,j=1}^{n-1}\begin{vmatrix} y_1^{l_1} & \ldots & y_n^{l_1}\\ \ldots & \ldots & \ldots\\ y_1^{l_{n-1}} & \ldots & y_n^{l_{n-1}}\\ 1 & \ldots & 1\end{vmatrix} = \frac{D(x_1,\ldots,x_{n-1})D(y_1,\ldots,y_n)\cdot x_1\ldots x_{n-1}}{\prod_{i=1}^{n-1}\prod_{j=1}^{n}(1-x_iy_j)}.$$

We change l_i to l_i+1. Then we have

$$\sum_{l_1>\cdots>l_{n-1}\geqslant 0}\det|x_i^{l_j}|_{i,j=1}^{n-1}\begin{vmatrix} y_1^{l_1+1} & \ldots & y_n^{l_1+1}\\ \ldots & \ldots & \ldots\\ y_1^{l_{n-1}+1} & \ldots & y_n^{l_{n-1}+1}\\ 1 & \ldots & 1\end{vmatrix} = \frac{D(x_1,\ldots,x_{n-1})D(y_1,\ldots,y_n)}{\prod_{i=1}^{n-1}\prod_{j=1}^{n}(1-x_iy_j)}.$$

Setting now $x_{n-1}=0$ and then repeating this process we arrive at the following result.

THEOREM 1.2.3. *Let $n \geqq m > 0$. For $|x_\nu| < 1, |y_\nu| < 1, \nu = 1, \cdots, n$, the following identity holds.*

$$\sum_{l_1 > \cdots > l_m \geqslant 0} \det \left| x_i^{l_j} \right|_{i,j=1}^{m} \begin{vmatrix} y_1^{l_1+n-m} & \cdots & y_n^{l_1+n-m} \\ \cdot & \cdot & \cdot \\ y_1^{l_m+n-m} & \cdots & y_n^{l_m+n-m} \\ y_1^{n-m-1} & \cdots & y_n^{n-m-1} \\ \cdot & \cdot & \cdot \\ y_1 & \cdots & y_n \\ 1 & & 1 \end{vmatrix}$$

$$= \frac{D(x_1, \ldots, x_m) D(y_1, \ldots, y_n)}{\prod_{i=1}^{m} \prod_{j=1}^{n} (1 - x_i y_j)}. \tag{1.2.6}$$

Later on we shall also need the following theorem.

THEOREM 1.2.4. *Let $f_1(x), \cdots, f_n(x)$ be sufficiently differentiable functions. Then*

$$\lim_{\substack{x_1 \to x \\ \cdots \\ x_n \to x}} \frac{\det |f_i(x_j)|_{i,j=1}^{n}}{D(x_1, \ldots, x_n)}$$

$$= \frac{(-1)^{n(n-1)/2}}{1!\,2! \ldots (n-1)!} \det \left| f_i^{(j-1)}(x) \right|_{i,j=1}^{n}. \tag{1.2.7}$$

PROOF. It is sufficient to consider the case where $x = 0$. If $f_i(x)$ are analytic functions, then our theorem is a consequence of Theorem 1.2.1; for non-analytic functions we have to repeat the same considerations using MacLaurin's formula with remainder term.

Let $f_1 \geqq f_2 \geqq \cdots \geqq f_n \geqq 0$ be integers. We introduce the notation

$$M_{f_1, \ldots, f_n}(x_1, \ldots, x_n) = \det \left| x_j^{f_i+n-i} \right|_{i,j=1}^{n}. \tag{1.2.8}$$

It is clear that

$$M_{0,0,\ldots,0}(x_1, \ldots, x_n) = D(x_1, \ldots, x_n).$$

We set

$$N(f_1, \ldots, f_n) = \lim_{\substack{x_1 \to 1 \\ \cdots \\ x_n \to 1}} \frac{M_{f_1, \ldots, f_n}(x_1, \ldots, x_n)}{D(x_1, \ldots, x_n)}. \tag{1.2.9}$$

From Theorem 1.2.4 we obtain

$$N(f_1, \ldots, f_n) = \frac{D(f_1+n-1, f_2+n-2, \ldots, f_{n-1}+1, f_n)}{D(n-1, n-2, \ldots, 1, 0)}. \qquad (1.2.10)$$

THEOREM 1.2.5. *Let $n \geqq m \geqq 1$. For $|x_i| < 1$, $i = 1, 2, \cdots, m$, we have*

$$\prod_{i=1}^{m} (1-x_i)^{-\rho-n+1} = C_\rho \sum_{l_1 > \cdots > l_m \geqslant 0} a_{l_1+n-m} \cdots a_{l_m+n-m}$$

$$\times N(f_1, \ldots, f_m, 0, \ldots, 0) \frac{M_{f_1, \ldots, f_m}(x_1, \ldots, x_m)}{D(x_1, \ldots, x_m)}, \qquad (1.2.11)$$

where

$$l_1 = f_1 + m - 1, \ldots, l_{m-1} = f_{m-1} + 1, \quad l_m = f_m,$$

$$a_l = \frac{\Gamma(\rho+l)}{\Gamma(\rho)\Gamma(l+1)}, \quad C_\rho = \frac{1}{a_{n-m} \cdots a_{n-1}}.$$

PROOF. (1) First let $m = n$. Setting

$$f(z) = (1-z)^{-\rho} = \sum_{l=0}^{\infty} a_l z^l,$$

in Theorem 1.2.2, for $|x_i y_j| < 1$ we obtain the identity

$$\det|(1-x_i y_j)^{-\rho}|_1^n = \sum_{l_1 > \cdots > l_n \geqslant 0} a_{l_1} \cdots a_{l_n} \det|x_i^{l_j}|_{i,j=1}^n \det|y_i^{l_j}|_{i,j=1}^n .$$

Dividing this identity by $D(y_1, \cdots, y_n)$ and letting $y_1 \to 1, \cdots, y_n \to 1$, by Theorem 1.2.4 we obtain (since $\rho(\rho+1) \cdots (\rho+j-2) = 1$ for $j = 1$),

$$\lim_{\substack{y_1 \to 1 \\ \cdots \\ y_n \to 1}} \frac{\det|(1-x_i y_j)^{-\rho}|_1^n}{D(y_1, \ldots, y_n)}$$

$$= \frac{(-1)^{\frac{n(n-1)}{2}}}{1!\,2! \ldots (n-1)!} \det|\rho(\rho+1) \ldots (\rho+j-2) x_i^{j-1} (1-x_i)^{-\rho-j+1}|_1^n$$

$$= \frac{\rho^{n-1}(\rho+1)^{n-2} \ldots (\rho+n-2)}{1!\,2! \ldots (n-1)!} D\left(\frac{x_1}{1-x_1}, \ldots, \frac{x_n}{1-x_n}\right)$$

$$\times \prod_{i=1}^{n} (1-x_i)^{-\rho} = a_0 a_1 \ldots a_{n-1} D(x_1, \ldots, x_n) \prod_{i=1}^{n} (1-x_i)^{-\rho-n+1}.$$

Whence it follows that

$$\prod_{i=1}^{n}(1-x_i)^{-\rho-n+1}=\frac{1}{a_0\ldots a_{n-1}}\sum_{l_1>\ldots>l_n\geqslant 0} a_{l_1}\ldots a_{l_n}$$
$$\times N(f_1,\ldots,f_n)\frac{M_{f_1,\ldots,f_n}(x_1,\ldots,x_n)}{D(x_1,\ldots,x_n)}. \quad (1.2.12)$$

(2) Now let $m<n$. We set $x_n=0$ in (1.2.12). Then all terms with $l_n>0$ vanish, and so

$$\prod_{i=1}^{n-1}(1-x_i)^{-\rho-n+1}=\frac{1}{a_0\ldots a_{n-1}}\sum_{l_1>\ldots>l_{n-1}>0} a_{l_1}\ldots a_{l_{n-1}}$$
$$\times N(f_1,\ldots,f_{n-1},0)\frac{M_{f_1,\ldots,f_{n-1}}(x_1,\ldots,x_{n-1})}{D(x_1,\ldots,x_{n-1})}.$$

Changing l_ν to $l_\nu+1$, we get

$$\prod_{i=1}^{n-1}(1-x_i)^{-\rho-n+1}=\frac{1}{a_0\ldots a_{n-1}}\sum_{l_1>\ldots>l_{n-1}\geqslant 0} a_{l_1+1}\ldots a_{l_{n-1}+1}$$
$$\times N(f_1,\ldots,f_{n-1},0)\frac{M_{f_1,\ldots,f_{n-1}}(x_1,\ldots,x_{n-1})}{D(x_1,\ldots,x_{n-1})}.$$

Setting then $x_{n-1}=0$ and repeating this process sufficiently many times we obtain the following theorem.

THEOREM 1.2.6. *For $|x|<1$, $|x_k|<1$, $k=1,\cdots,n$, the following identity holds.*

$$\sum_{f_1\geqslant\ldots\geqslant f_n\geqslant 0} M_{2f_1,\ldots,2f_n}(x_1,\ldots,x_n)\,x^{f_1+\cdots+f_n}$$
$$=D(x_1,\ldots,x_n)\prod_{1\leqslant i\leqslant j\leqslant n}(1-xx_ix_j)^{-1}.$$

PROOF. The left-hand side of this identity is equal to

$$\sum_{f_1\geqslant\ldots\geqslant f_n\geqslant 0}\ \sum_{i_1,\ldots,i_n}\delta^{1\ldots n}_{i_1\ldots i_n}x_{i_1}^{2f_1+n-1}x_{i_2}^{2f_2+n-2}\ldots x_{i_n}^{2f_n}\cdot x^{f_1+\cdots+f_n}$$
$$=\sum_{i_1,\ldots,i_n}\delta^{1\ldots n}_{i_1\ldots i_n}x_{i_1}^{n-1}x_{i_2}^{n-2}\ldots x_{i_{n-1}}\sum_{f_1\geqslant\ldots\geqslant f_n\geqslant 0}\left(xx_{i_1}^2\right)^{f_1}\ldots\left(xx_{i_n}^2\right)^{f_n}$$
$$=\sum_{i_1,\ldots,i_n}\delta^{1\ldots n}_{i_1\ldots i_n}\frac{x_{i_1}^{n-1}x_{i_2}^{n-2}\ldots x_{i_{n-1}}}{\left(1-xx_{i_1}^2\right)\left(1-x^2x_{i_1}^2x_{i_2}^2\right)\ldots\left(1-x^nx_{i_1}^2\ldots x_{i_n}^2\right)}. \quad (1.2.13)$$

Here we used the following trivial identity: for $|a_\nu|<1$, $\nu=1,2,\cdots,n$,

$$\sum_{f_1\geqslant\cdots\geqslant f_n\geqslant 0} a_1^{f_1}\cdots a_n^{f_n} = (1-a_1)^{-1}(1-a_1a_2)^{-1}\cdots(1-a_1a_2\cdots a_n)^{-1}. \qquad (1.2.14)$$

Substituting $x_i\sqrt{x}$ for x_i in Theorem 1.1.1, we have

$$\sum_{i_1,\ldots,i_n} \delta_{i_1\cdots i_n}^{1\cdots n} \frac{x_{i_1}^{n-1}x_{i_2}^{n-2}\cdots x_{i_{n-1}}(\sqrt{x})^{\frac{n(n-1)}{2}}}{\left(1-xx_{i_1}^2\right)\cdots\left(1-x^n x_{i_1}^2\cdots x_{i_n}^2\right)} = (\sqrt{x})^{\frac{n(n-1)}{2}} D(x_1,\ldots,x_n)\prod_{1\leqslant i\leqslant j\leqslant n}(1-xx_ix_j)^{-1}. \qquad (1.2.15)$$

Substituting this expression into (1.2.13) we get the statement of the theorem.

Theorem 1.2.7. *Let* $\nu=[n/2]$. *For* $|x|<1, |x_k|<1, k=1,\cdots,n$, *the identity*

$$\sum_{f_1\geqslant\cdots\geqslant f_\nu\geqslant 0} M_{f_1,f_1,f_2,f_2,\ldots}(x_1,\ldots,x_n)\,x^{f_1+\cdots+f_\nu} = D(x_1,\ldots,x_n)\prod_{1\leqslant i<j\leqslant n}(1-xx_ix_j)^{-1}, \qquad (1.2.16)$$

holds (*for odd* n *the last index in* $M_{f_1,f_1,f_2,f_2,\cdots}$ *is* 0).

Proof. The left-hand side of (1.2.16) is equal to

$$\sum_{f_1\geqslant\cdots\geqslant f_\nu\geqslant 0}\ \sum_{i_1,\ldots,i_n} \delta_{i_1\cdots i_n}^{1\cdots n} x_{i_1}^{f_1+n-1}x_{i_2}^{f_1+n-2}\cdots x^{f_1+\cdots+f_\nu}$$

$$= \sum_{i_1,\ldots,i_n}\delta_{i_1\cdots i_n}^{1\cdots n} x_{i_1}^{n-1}x_{i_2}^{n-2}\cdots x_{i_{n-1}} \sum_{f_1\geqslant\cdots\geqslant f_\nu\geqslant 0}(xx_{i_1}x_{i_2})^{f_1}(xx_{i_3}x_{i_4})^{f_2}\cdots$$

$$= \sum_{i_1,\ldots,i_n}\delta_{i_1\cdots i_n}^{1\cdots n} x_{i_1}^{n-1}x_{i_2}^{n-2}\cdots x_{i_{n-1}}(1-xx_{i_1}x_{i_2})^{-1}(1-x^2x_{i_1}x_{i_2}x_{i_3}x_{i_4})^{-1}\cdots.$$

By Theorem 1.1.2 this equals

$$D(x_1,\ldots,x_n)\prod_{1\leqslant i<j\leqslant n}(1-xx_ix_j)^{-1},$$

which was to be proved.

Theorem 1.2.8. *For* $|x|<1, |x_k|<1, k=1,2,\cdots,n$, *we have*

$$\sum_{f=0}^{\infty} M_{f,0,\ldots,0}(x_1,\ldots,x_n)\,x^f = D(x_1,\ldots,x_n)\prod_{i=1}^{n}(1-xx_i)^{-1}.$$

PROOF. By some simple transformations we find

$$\sum_{f=0}^{\infty} M_{f,0,\ldots,0}(x_1,\ldots,x_n)\,x^f$$

$$= \sum_{f=0}^{\infty}\begin{vmatrix} (x_1x)^f x_1^{n-1} & x_1^{n-2} & \ldots & x_1 & 1 \\ \cdot & \cdot & \cdot & \cdot & \cdot \\ (x_nx)^f x_n^{n-1} & x_n^{n-2} & \ldots & x_n & 1 \end{vmatrix} = \begin{vmatrix} \dfrac{x_1^{n-1}}{1-xx_1} & x_1^{n-2} & \ldots & x_1 & 1 \\ \cdot & \cdot & \cdot & \cdot & \cdot \\ \dfrac{x_n^{n-1}}{1-xx_n} & x_n^{n-2} & \ldots & x_n & 1 \end{vmatrix}$$

$$= D(x_1,\ldots,x_n)\prod_{i=1}^{n}(1-xx_i)^{-1}.$$

THEOREM 1.2.9.

$$\sum_{l_1>\cdots>l_n\geqslant 0}\frac{[l_1+\ldots+l_n-n(n-1)/2]!}{l_1!\,l_2!\ldots l_n!}D(l_1,\ldots,l_n)\det\left|x_i^{l_j}\right|_{i,j=1}^{n}$$

$$=\frac{D(x_1,\ldots,x_n)}{1-x_1-\ldots-x_n}.$$

PROOF. Setting in Theorem 1.2.2 $f(z)=e^z=\sum_{l=0}^{\infty}(z^l/l!)$ we get

$$\sum_{l_1>\cdots>l_n\geqslant 0}\frac{1}{l_1!\ldots l_n!}\det\left|y_i^{l_j}\right|_{i,j=1}^{n}\det\left|x_i^{l_j}\right|_{i,j=1}^{n}=\det\left|e^{x_iy_j}\right|_1^n.$$

Since

$$\left[l_1+\ldots+l_n-\frac{n(n-1)}{2}\right]!=\int_0^{\infty}e^{-t}\,t^{l_1+\cdots+l_n-n(n-1)/2}\,dt,$$

it follows that

$$\sum_{l_1>\cdots>l_n\geqslant 0}\frac{[l_1+\ldots+l_n-n(n-1)/2]!}{l_1!\ldots l_n!}\det\left|y_i^{l_j}\right|_{i,j=1}^{n}\det\left|x_i^{l_j}\right|_{i,j=1}^{n}$$

$$=\int_0^{\infty}t^{-\frac{n(n-1)}{2}}e^{-t}\sum_{l_1>\cdots>l_n\geqslant 0}\frac{1}{l_1!\ldots l_n!}\det\left|y_i^{l_j}\right|_{i,j=1}^{n}\det\left|(tx_i)^{l_j}\right|_{i,j=1}^{n}dt$$

$$=\int_0^{\infty}t^{-\frac{n(n-1)}{2}}e^{-t}\det\left|e^{tx_iy_j}\right|_1^n dt.$$

Since by Theorem 1.2.4

$$\lim_{\substack{y_1 \to 1 \\ \cdots \\ y_n \to 1}} \frac{\det \left| y_i^{l_j} \right|_{i,j=1}^n}{D(y_1, \ldots, y_n)} = \frac{D(l_1, l_2, \ldots, l_n)}{1!\, 2! \ldots (n-1)!},$$

we obtain, using Theorem 1.2.4,

$$\sum_{l_1 > \cdots > l_n \geqslant 0} \frac{[l_1 + \ldots + l_n - n(n-1)/2]!}{l_1! \ldots l_n!} D(l_1, \ldots, l_n) \det \left| x_i^{l_j} \right|_{i,l=1}^n$$

$$= (-1)^{\frac{n(n-1)}{2}} \int_0^\infty t^{-\frac{n(n-1)}{2}} e^{-t} \det \left| e^{x_i t} (x_i t)^{j-1} \right|_1^n dt$$

$$= D(x_1, \ldots, x_n) \int_0^\infty e^{-t + (x_1 + \ldots + x_n) t} dt = \frac{D(x_1, \ldots, x_n)}{1 - x_1 - \ldots - x_n},$$

which was to be proved.

Setting $x_1 \to x, \cdots, x_n \to x$, the following theorem follows.

THEOREM 1.2.10.

$$\sum_{l_1=0}^\infty \cdots \sum_{l_n=0}^\infty \frac{[l_1 + \ldots + l_n - n(n-1)/2]!}{l_1! \ldots l_n!} D^2(l_1, \ldots, l_n) x^{l_1 + \cdots + l_n}$$

$$= 1!\, 2! \ldots n! \frac{x^{\frac{n(n-1)}{2}}}{1 - nx}.$$

Comparing the coefficients of the powers of x we obtain

$$\sum_{l_1 + \ldots + l_n = m} \frac{D^2(l_1, \ldots, l_n)}{l_1! \ldots l_n!} = \frac{n^{m - \frac{n(n-1)}{2}} \cdot 1!\, 2! \ldots n!}{\left(m - \frac{n(n-1)}{2}\right)!}. \qquad (1.2.17)$$

1.3. **Identities for** $N(f_1, \cdots, f_n)$. In this section we prove some identities involving $N(f_1, \cdots, f_n)$. These identities are special cases of those in §1.4. However, starting with these we can see better how to obtain the identities of §1.4.

THEOREM 1.3.1. *Let* $n \geqq m > 0$. *For* $|x| < 1$ *we have*

$$\sum_{f_1 \geqslant \cdots \geqslant f_m \geqslant 0} N(f_1, \ldots, f_m) N(f_1, \ldots, f_m, 0, \ldots, 0) x^{f_1 + \cdots + f_m} = (1-x)^{-mn}, \qquad (1.3.1)$$

whence by comparing coefficients we obtain

$$\sum_{\substack{f_1 + \cdots + f_m = f \\ f_1 \geqslant \cdots \geqslant f_m \geqslant 0}} N(f_1, \ldots, f_m) N(f_1, \ldots, f_m, 0, \ldots, 0) = \frac{(mn+f-1)!}{(mn-1)!\,f!}. \qquad (1.3.2)$$

Proof. Changing x_i to xx_i in Theorem 1.2.3 and letting $x_i \to 1$, $y_i \to 1$, we get (1.3.1).

Theorem 1.3.2. *For* $|x| < 1$

$$\sum_{f_1 \geqslant \cdots \geqslant f_n \geqslant 0} N(2f_1, \ldots, 2f_n) x^{f_1 + \cdots + f_n} = (1-x)^{-\frac{n(n+1)}{2}}, \qquad (1.3.3)$$

whence by comparing coefficients we obtain

$$\sum_{\substack{f_1 + \cdots + f_n = f \\ f_1 \geqslant \cdots \geqslant f_n \geqslant 0}} N(2f_1, \ldots, 2f_n) = \frac{[f + n(n+1)/2 - 1]!}{f!\,[n(n+1)/2 - 1]!}.$$

Proof. (1.3.3) follows easily from Theorem 1.2.6.

Theorem 1.3.3. *For* $|x| < 1$

$$\sum_{f_1 \geqslant \cdots \geqslant f_\nu \geqslant 0} N(f_1, f_1, f_2, f_2, \ldots) x^{f_1 + \cdots + f_\nu} = (1-x)^{-\frac{n(n-1)}{2}}, \qquad (1.3.4)$$

whence by comparing coefficients we obtain

$$\sum_{\substack{f_1 \geqslant \cdots \geqslant f_\nu \geqslant 0 \\ f_1 + \cdots + f_\nu = f}} N(f_1, f_1, f_2, f_2, \ldots) = \frac{[f + n(n-1)/2 - 1]!}{f!\,[n(n-1)/2 - 1]!}.[8] \qquad (1.3.5)$$

---

[8] Translator's note. (1.3.4) follows from Theorem 1.2.7.

1.4. **Identities for characters.** Let $f_1 \geqq f_2 \geqq \cdots \geqq f_n \geqq 0$ be integers and let

$$A_{f_1, \cdots, f_n}(X) \tag{1.4.1}$$

be a representation of the full linear group GL(n) (i.e., the group of all nonsingular linear transformations of complex n-space) with signature $(f_1, f_2, \cdots, f_n)$.[9] We assume that this representation is unitary for unitary X. As is well known, (1.4.1) is a matrix with $N(f_1, \cdots, f_n)$ rows and columns. The trace of this matrix is denoted by

$$\chi_{f_1, \ldots, f_n}(X) = \mathrm{Sp}\, A_{f_1, \ldots, f_n}(X).$$

This quantity is called the character of the representation (1.4.1).

Suppose that X is a diagonal matrix, $X = \Lambda = [\lambda_1, \cdots, \lambda_n]$. Then, as is well known,[10]

$$\chi_{f_1, \ldots, f_n}(\Lambda) = \frac{M_{f_1, \ldots, f_n}(\lambda_1, \ldots, \lambda_n)}{D(\lambda_1, \ldots, \lambda_n)}. \tag{1.4.2}$$

Now we introduce the following two abbreviated notations:

$$A_{f, 0, \ldots, 0}(X) = A^{[f]}(X) = X^{[f]},$$

$$A_{\underbrace{1, \ldots, 1}_{f \text{ times}}, 0, \ldots, 0}(X) = A^{(f)}(X) = X^{(f)}.$$

THEOREM 1.4.1. *Let $n \geqq m > 0$, and let X be an element of the full linear group* GL(m), *Y an element of the group* GL(n). *We denote by $X \times Y$ the Kronecker product of X and Y. Then*

$$\mathrm{Sp}((X \times Y)^{[f]}) = \sum_{\substack{f_1 + \cdots + f_m = f \\ f_1 \geqslant \cdots \geqslant f_m \geqslant 0}} \chi_{f_1, \ldots, f_m}(X)\, \chi_{f_1, \ldots, f_m, 0, \ldots, 0}(Y). \tag{1.4.3}$$

PROOF. First we consider the special case where X and Y are diagonal matrices

$$X = [x_1, \ldots, x_m], \quad Y = [y_1, \ldots, y_n].$$

Changing x_i to xx_i in Theorem 1.2.3, we have by this theorem

[9] Translator's note. Cf. H. Weyl [**2**, p. 181].

[10] Translator's note. Cf. H. Weyl [**2**, pp. 276 ff.].

$$\sum_{f_1 \geqslant \cdots \geqslant f_m \geqslant 0} \chi_{f_1, \cdots, f_m}(X) \chi_{f_1, \cdots, f_m, 0, \cdots, 0}(Y) x^{f_1+\cdots+f_m} = \prod_{i=1}^{m} \prod_{j=1}^{n} (1 - x x_i y_j)^{-1}. \quad (1.4.4)$$

On the other hand, by Theorem 1.2.8 we have

$$\sum_{f=0}^{\infty} \mathrm{Sp}\,(X \times Y)^{[f]} x^f = \prod_{i=1}^{m} \prod_{j=1}^{n} (1 - x x_i y_j)^{-1}. \quad (1.4.5)$$

Comparing the coefficients of x^f in (1.4.4) and (1.4.5) we obtain the statement of the theorem for our special case of diagonal X and Y.

Since the terms of the series on the right-hand side of (1.4.3) remain unchanged when we change X and Y to PXP^{-1} and QYQ^{-1}, it follows that (1.4.3) remains true for any matrices X and Y that can be diagonalized by a similarity transformation. By reasons of continuity the theorem is therefore true in the general case too.

THEOREM 1.4.2. *The following identity holds.*

$$\sum_{\substack{f_1+\cdots+f_n=f \\ f_1 \geqslant \cdots \geqslant f_n \geqslant 0}} \chi_{2f_1, \cdots, 2f_n}(X) = \mathrm{Sp}\,(X^{[2]})^{[f]}. \quad (1.4.6)$$

PROOF. First suppose that $X=[x_1, \cdots, x_n]$. Then, by Theorem 1.2.6 we have

$$\sum_{f_1 \geqslant \cdots \geqslant f_n \geqslant 0} \chi_{2f_1, \cdots, 2f_n}(X) x^{f_1+\cdots+f_n} = \prod_{1 \leqslant i \leqslant j \leqslant n} (1 - x x_i x_j)^{-1}.$$

On the other hand, substituting in Theorem 1.2.8 $n(n+1)/2$ and $x_i x_j$, $i \leqq j$, $i, j=1, \cdots, n$ for n and $x_1, \cdots, x_n$, we obtain

$$\sum_{f=0}^{\infty} \mathrm{Sp}\,((X^{[2]})^{[f]}) x^f = \prod_{1 \leqslant i \leqslant j \leqslant n} (1 - x x_i x_j)^{-1}.$$

Comparing coefficients and using the same relation as in the proof of the preceding theorem, we obtain our assertion.

Finally, the following theorem can be proved in an analogous manner.

THEOREM 1.4.3.

$$\sum_{\substack{f_1+\cdots+f_\nu=f \\ f_1 \geqslant \cdots \geqslant f_\nu \geqslant 0}} \chi_{f_1, f_1, f_2, f_2, \cdots}(X) = \mathrm{Sp}\,(X^{(2)})^{[f]}.$$

CHAPTER II

EVALUATION OF SOME INTEGRALS

2.1. Matrix analogues of the integral $\int_{-\infty}^{+\infty}(1+x^2)^{-\alpha}dx$.

THEOREM 2.1.1. *If $\alpha > n/2$, then*

$$I_n(\alpha) = \int_T \frac{\dot{T}}{(\det(I+T^2))^\alpha}$$

$$= 2^{\frac{n(n-1)}{4}} \pi^{\frac{n(n+1)}{4}} \frac{\Gamma\left(\alpha - \frac{n}{2}\right)}{\Gamma(\alpha)} \prod_{\nu=1}^{n-1} \frac{\Gamma\left(2\alpha - \frac{n+\nu}{2}\right)}{\Gamma(2\alpha-\nu)}. \quad (2.1.1)$$

Here $T=(t_{jk})_1^n$ varies over the set of all real symmetric matrices of order n, and $\dot{T}=2^{n(n-1)/4}\prod_{j\leq k} dt_{jk}$.

As a preliminary, we prove the following two theorems.

THEOREM 2.1.2. *Let Z be an $m\times n$ matrix. Then*

$$\det(I^{(m)} - Z\bar{Z}') = \det(I^{(n)} - \bar{Z}'Z). \quad (2.1.2)$$

Furthermore, the conditions $I^{(m)} - Z\bar{Z}' > 0$ and $I^{(n)} - \bar{Z}'Z > 0$ are equivalent. ($I^{(m)}$ denotes the identity matrix of order m.)

PROOF. It is well known that any $m\times n$ matrix Z can be represented in the form

$$Z = U\Lambda V,$$

where U and V are unitary matrices of order m and n, respectively, and the $m\times n$ matrix Λ is of the form

$$\Lambda = \begin{pmatrix} \lambda_1 & 0 & \dots & 0 \\ 0 & \lambda_2 & \dots & 0 \\ \dots & \dots & \dots & \dots \end{pmatrix}, \quad \lambda_\nu \geq 0.$$

It follows that

$$\det(I^{(m)} - Z\bar{Z}') = (1-\lambda_1^2)(1-\lambda_2^2)\ldots = \det(I^{(n)} - \bar{Z}'Z).$$

The second assertion of the theorem follows from the fact that both of the conditions mentioned there are equivalent to $|\lambda_1|<1$, $|\lambda_2|<1, \cdots$.

THEOREM 2.1.3. *Let* $a>0$, $b^2-ac<0$, $\alpha>\frac{1}{2}$. *Then*

$$\int_{-\infty}^{\infty} \frac{dx}{(ax^2+2bx+c)^\alpha} = a^{\alpha-1}(ac-b^2)^{\frac{1}{2}-\alpha}\sqrt{\pi}\,\frac{\Gamma\left(\alpha-\frac{1}{2}\right)}{\Gamma(\alpha)}. \tag{2.1.3}$$

PROOF. We make the substitution $y=(a/\sqrt{(ac-b^2)})\,(x+b/a)$. Then

$$ax^2+bx+c = \frac{ac-b^2}{a}(y^2+1), \quad dx = \frac{\sqrt{ac-b^2}}{a}\,dy\,.$$

Therefore

$$\int_{-\infty}^{\infty} \frac{dx}{(ax^2+2bx+c)^\alpha} = a^{\alpha-1}(ac-b^2)^{\frac{1}{2}-\alpha}\int_{-\infty}^{\infty}\frac{dy}{(1+y^2)^\alpha}$$

$$= a^{\alpha-1}(ac-b^2)^{\frac{1}{2}-\alpha}\sqrt{\pi}\,\frac{\Gamma\left(\alpha-\frac{1}{2}\right)}{\Gamma(\alpha)}.$$

PROOF OF THEOREM 2.1.1. We set

$$T = \begin{pmatrix} T_1 & v' \\ v & t \end{pmatrix} \qquad (t=t_{nn}),$$

where T_1 is a real symmetric matrix of order $n-1$, v is an $(n-1)$-dimensional vector, and t is a real number. Then

$$I+T^2 = \begin{pmatrix} I+T_1^2+v'v & T_1v'+v't \\ vT_1+tv & 1+vv'+t^2 \end{pmatrix}.$$

Since for $A=A'$,

$$\begin{pmatrix} I & 0 \\ -bA^{-1} & 1 \end{pmatrix}\begin{pmatrix} A & b' \\ b & c \end{pmatrix}\begin{pmatrix} I & 0 \\ -bA^{-1} & 1 \end{pmatrix}' = \begin{pmatrix} A & 0 \\ 0 & c-b\,A^{-1}b' \end{pmatrix}, \tag{2.1.4}$$

it follows that

$$\det(I+T^2) = \det(I+T_1^2+v'v)$$
$$\times\{1+vv'+t^2-(vT_1+tv)(I+T_1^2+v'v)^{-1}(T_1v'+v't)\}.$$

The second factor on the right-hand side of this equality can be written in the form $at^2+2bt+c$, where

$$a = 1 - v(I + T_1^2 + v'v)^{-1}v',$$
$$2b = -vT_1(I + T_1^2 + v'v)^{-1}v' - v(I + T_1^2 + v'v)^{-1}T_1v'$$
$$= -2v(I + T_1^2 + v'v)^{-1}T_1v',$$
$$c = 1 + vv' - vT_1(I + T_1^2 + v'v)^{-1}T_1v'.$$

The symmetric matrix T_1 can be written in the form

$$T_1 = \Gamma'[\lambda_1, \ldots, \lambda_{n-1}]\Gamma,$$

where Γ is an orthogonal matrix.

We set

$$T_2 = \Gamma'\left[\sqrt{1+\lambda_1^2}, \ldots, \sqrt{1+\lambda_{n-1}^2}\right]\Gamma.$$

Then

$$T_2 = T_2', \quad T_1T_2 = T_2T_1, \quad I + T_1^2 = T_2^2.$$

If we set $v = wT_2$, we obtain

$$\dot{v} = \det T_2 \cdot \dot{w} = \left(\det\left(I + T_1^2\right)\right)^{\frac{1}{2}} \cdot \dot{w}$$

and

$$I + T_1^2 + v'v = T_2(I + w'w)T_2.$$

Besides, if u is an $(n-1)$-dimensional vector, then [by the equality $(w'w)^2 = w'w(ww')$] we have

$$u(I + w'w)^{-1}u' = uu' - \frac{(uw')^2}{1 + ww'}$$

and

$$w(I + w'w)^{-1} = \frac{w}{1 + ww'}.$$

That is

$$a = 1 - w(I + w'w)^{-1}w' = \frac{1}{1 + ww'},$$

$$b = -w(I + w'w)^{-1}T_1w' = -\frac{wT_1w'}{1 + ww'},$$

$$c = 1 + wT_2^2w' - wT_1(I + w'w)^{-1}T_1w' = 1 + ww' + \frac{(wT_1w')^2}{1 + ww'}.$$

Consequently,

$$ac - b^2 = 1.$$

By Theorem 2.1.3 we have

$$I_n(\alpha) = \int_T \frac{\dot{T}}{[\det(I+T^2)]^\alpha}$$

$$= 2^{\frac{n-1}{2}} \int_{t,v,T_1} [\det(I+T_1^2+v'v)]^{-\alpha} (at^2+2bt+c)^{-\alpha} dt\,\dot{v}\,\dot{T}_1$$

$$= 2^{\frac{n-1}{2}} \sqrt{\pi}\,\frac{\Gamma\left(\alpha-\frac{1}{2}\right)}{\Gamma(\alpha)} \int_w (1+ww')^{1-2\alpha}\dot{w} \int_{T_1} [\det(I+T_1^2)]^{\frac{1}{2}-\alpha}\dot{T}_1.$$

Making use of the formula $(\alpha > (n+1)/4)$

$$\int_{-\infty}^{\infty} \cdots \int_{-\infty}^{\infty} (1+x_1^2+\ldots+x_{n-1}^2)^{1-2\alpha} dx_1 \ldots dx_{n-1}$$

$$= \pi^{\frac{n-1}{2}} \frac{\Gamma\left(2\alpha-\frac{n+1}{2}\right)}{\Gamma(2\alpha-1)}, \qquad (2.1.5)$$

we get the recursive relation

$$I_n(\alpha) = 2^{\frac{n-1}{2}} \pi^{\frac{n}{2}} \frac{\Gamma\left(2\alpha-\frac{n+1}{2}\right)\Gamma\left(\alpha-\frac{1}{2}\right)}{\Gamma(\alpha)\Gamma(2\alpha-1)} I_{n-1}\left(\alpha-\frac{1}{2}\right). \qquad (2.1.6)$$

Applying formula (2.1.6) $n-1$ times and noting that

$$I_1\left(\alpha-\frac{n-1}{2}\right) = \int_{-\infty}^{\infty} \frac{dx}{(1+x^2)^{\alpha-\frac{n-1}{2}}} = \frac{\sqrt{\pi}\,\Gamma\left(\alpha-\frac{n}{2}\right)}{\Gamma\left(\alpha-\frac{n-1}{2}\right)} \qquad \left(\alpha > \frac{n}{2}\right),$$

we get the assertion.

THEOREM 2.1.4. *Let* $n \geqq 2$. *For* $\alpha > (2n-3)/4$

$$J_n(\alpha) = \int_K \frac{\dot{K}}{(\det(1+KK'))^\alpha} = 2^{\frac{n(n-1)}{4}} \pi^{\frac{n(n-1)}{4}} \prod_{\nu=2}^{n} \frac{\Gamma\left(2\alpha-n+\frac{\nu+1}{2}\right)}{\Gamma(2\alpha-n+\nu)}, \qquad (2.1.7)$$

where K *varies over all real skew-symmetric matrices of order* n, *and* $\dot{K} = 2^{n(n-1)/4} \prod_{i<j} dk_{ij}$.

PROOF. We write K in the form

$$K=\begin{pmatrix} K_1 & -v' \\ v & 0 \end{pmatrix},$$

where K_1 is an $(n-1)$-dimensional real skew-symmetric matrix, and v is an $(n-1)$-dimensional vector. Then

$$I+KK'=\begin{pmatrix} I+K_1K_1'+v'v & K_1v' \\ vK_1' & 1+vv' \end{pmatrix}.$$

Using a formula analogous to (2.1.4), we obtain

$$\det(I+KK') \\ =\{1+vv'-vK_1'(I+K_1K_1'+v'v)^{-1}K_1v'\}\det(I+K_1K_1'+v'v).$$

We can find an orthogonal matrix Γ such that

$$K_1=\Gamma\left\{\begin{pmatrix} 0 & \lambda_1 \\ -\lambda_1 & 0 \end{pmatrix}\dotplus\begin{pmatrix} 0 & \lambda_2 \\ -\lambda_2 & 0 \end{pmatrix}\dotplus\cdots\right\}\Gamma',$$

where the last term in braces equals $\begin{pmatrix} 0 & \lambda_{n/2} \\ -\lambda_{n/2} & 0 \end{pmatrix}$, if n is even, and 0 if n is odd.

We set

$$T=\Gamma[\sqrt{1+\lambda_1^2},\ \sqrt{1+\lambda_1^2},\ \sqrt{1+\lambda_2^2},\ \sqrt{1+\lambda_2^2},\ldots]\Gamma'.$$

Then

$$T=T',\quad K_1T=TK_1,\quad I+K_1K_1'=T^2.$$

Substituting $v=wT$. We have

$$\dot{v}=\dot{w}\det T=\dot{w}\,[\det(I+K_1K_1')]^{\frac{1}{2}},$$

$$I+K_1K_1'+v'v=T^2+Tw'wT=T(I+w'w)T.$$

Therefore

$$\begin{aligned} &1+vv'-vK_1'(I+K_1K_1'+v'v)^{-1}K_1v' \\ &=1+wT^2w'-wTK_1'T^{-1}(I+w'w)^{-1}T^{-1}K_1Tw' \\ &=1+wT^2w'-wK_1'(I+w'w)^{-1}K_1w' \\ &=1+ww'-\frac{(wK_1w')^2}{1+ww'}=1+ww'. \end{aligned}$$

Consequently,

$$J_n(\alpha) = \int_K (\det(I + KK'))^{-\alpha} \dot{K}$$

$$= 2^{\frac{n-1}{2}} \int_{K_1} (\det(I + K_1K_1')^{\frac{1}{2}-\alpha} \dot{K}_1 \cdot \int_w (1 + ww')^{-2\alpha} \dot{w}$$

$$= 2^{\frac{n-1}{2}} \pi^{\frac{n-1}{2}} \frac{\Gamma\left(2\alpha - \frac{n-1}{2}\right)}{\Gamma(2\alpha)} J_{n-1}\left(\alpha - \frac{1}{2}\right).$$

Repeated application of this formula leads us to the integral

$$J_2\left(\alpha - \frac{n-2}{2}\right) = \sqrt{2} \int_{-\infty}^{\infty} (1 + t^2)^{n-2\alpha-2} dt$$

$$= \sqrt{2}\, \pi^{\frac{1}{2}} \frac{\Gamma\left(2\alpha - n + \frac{3}{2}\right)}{\Gamma(2\alpha - n + 2)} \qquad \left(\alpha > \frac{2n-3}{4}\right),$$

and we obtain the assertion of the theorem.

THEOREM 2.1.5. *Let* $\alpha > n - \frac{1}{2}$. *Then*

$$H_n(\alpha) = \int_H (\det(I + H^2))^{-\alpha} \dot{H}$$

$$= 2^{\frac{n(n-1)}{2}} \pi^{\frac{n^2}{2}} \prod_{j=0}^{n-1} \frac{\Gamma\left(\alpha - j - \frac{1}{2}\right)}{\Gamma(\alpha - j)} \cdot \prod_{k=0}^{n-2} \frac{\Gamma(2\alpha - n - k)}{\Gamma(2\alpha - 2k - 1)}. \qquad (2.1.8)$$

Here $H = (h_{jk})_1^n$ *varies over all hermitian matrices,*

$$h_{jj} = h_j, \quad h_{jk} = h'_{jk} + ih''_{jk}, \quad j < k, \quad \dot{H} = 2^{\frac{n(n-1)}{2}} \prod_{j=1}^{n} dh_j \prod_{j<k} dh'_{jk} dh''_{jk}.$$

PROOF. We set

$$H = \begin{pmatrix} H_1 & \bar{v}' \\ v & h \end{pmatrix}, \quad (h = h_n),$$

where H_1 is a hermitian matrix of order $n-1$, v is an $(n-1)$-dimensional vector, h is a real number. Then

$$I + H^2 = \begin{pmatrix} I + H_1^2 + \bar{v}'v & H_1\bar{v}' + \bar{v}'h \\ vH_1 + hv & 1 + h^2 + v\bar{v}' \end{pmatrix}.$$

From the identity

$$\begin{pmatrix} I & 0 \\ -pA^{-1} & 1\end{pmatrix}\begin{pmatrix} A & \bar p' \\ p & l\end{pmatrix}\overline{\begin{pmatrix} I & 0 \\ -pA^{-1} & 1\end{pmatrix}}' = \begin{pmatrix} A & 0 \\ 0 & l - pA^{-1}\bar p'\end{pmatrix}, \tag{2.1.9}$$

valid for the hermitian matrix A, we obtain

$$\det(I+H^2) = (ah^2+2bh+c)\det\left(I+H_1^2+\bar v' v\right),$$

where

$$a = 1 - v\left(I+H_1^2+\bar v'v\right)^{-1}\bar v',$$

$$2b = -vH_1\left(I+H_1^2+\bar v'v\right)^{-1}\bar v' - v\left(I+H_1^2+\bar v'v\right)^{-1}H_1\bar v',$$

$$c = 1 + v\bar v' - vH_1\left(I+H_1^2+\bar v'v\right)^{-1}H_1\bar v'.$$

Since H_1 is a hermitian matrix, there exists a unitary matrix U such that

$$H_1 = U[\lambda_1, \ldots, \lambda_{n-1}]\bar U'.$$

We set

$$T = U\left[\sqrt{1+\lambda_1^2}, \ldots, \sqrt{1+\lambda_n^2}\right]\bar U'.$$

Then

$$T = \bar T', \quad TH_1 = H_1T, \quad I + H_1^2 = T^2.$$

If we make the substitution $v = uT$, then

$$\dot v = (\det T)^2\,\dot u = \det\left(I+H_1^2\right)\dot u,$$

$$I + H_1^2 + \bar v'v = T\left(I+\bar u'u\right)T.$$

Furthermore, since

$$\left(I+\bar u'u\right)^{-1}\bar u' = \frac{\bar u'}{1+u\bar u'}, \quad w\left(1+\bar u'u\right)^{-1}\bar w' = w\bar w' - \frac{|w\bar u'|^2}{1+u\bar u'},$$

(w is an arbitrary $(n-1)$-dimensional vector), we obtain

$$a = 1 - u(I+\bar u'u)^{-1}\bar u' = \frac{1}{1+u\bar u'} \quad (a>0),$$

$$b = -uH_1(I+\bar u'u)^{-1}\bar u' = -\frac{uH_1\bar u'}{1+u\bar u'},$$

$$c = 1 + uT^2\bar u' - uH_1(I+\bar u'u)^{-1}H_1\bar u' = 1 + u\bar u' + \frac{|uH_1\bar u'|^2}{1+u\bar u'}.$$

Since $uH_1\bar{u}'$ is a real number, we have $ac-b^2=1$. By Theorem 2.1.3 we have

$$H_n(\alpha)=\int_H [\det(I+H^2)]^{-\alpha}\dot{H}$$

$$=2^{n-1}\int_{u,\,H_1}\left[\det\left(I+H_1^2\right)\right]^{1-\alpha}(1+u\bar{u}')^{-\alpha}\,\dot{u}\dot{H}_1\cdot\int_{-\infty}^{\infty}(ah^2+2bh+c)^{-\alpha}\,dh$$

$$=2^{n-1}\pi^{\frac{1}{2}}\frac{\Gamma\left(\alpha-\frac{1}{2}\right)}{\Gamma(\alpha)}\int_u(1+u\bar{u}')^{1-2\alpha}\,\dot{u}\cdot\int_{H_1}\left[\det\left(I+H_1^2\right)\right]^{1-\alpha}\dot{H}_1$$

$$=2^{n-1}\pi^{n-\frac{1}{2}}\frac{\Gamma\left(\alpha-\frac{1}{2}\right)\Gamma(2\alpha-n)}{\Gamma(\alpha)\Gamma(2\alpha-1)}H_{n-1}(\alpha-1).$$

Applying this formula repeatedly and using the fact that

$$H_1(\alpha-n+1)=\int_{-\infty}^{\infty}(1+x^2)^{n-\alpha-1}\,dx=\pi^{\frac{1}{2}}\frac{\Gamma\left(\alpha-n+\frac{1}{2}\right)}{\Gamma(\alpha-n+1)}$$

$$\left(\alpha>n-\frac{1}{2}\right),$$

we get the assertion of the theorem.

2.2. The volume of $\mathfrak{R}_{\mathrm{I}}$.

THEOREM 2.2.1. *Let Z be an $m\times n$ matrix and $\lambda>-1$. We set*

$$J_{m,n}(\lambda)=\int_{I-Z\bar{Z}'>0}[\det(I-Z\bar{Z}')]^{\lambda}\dot{Z},$$

where $\dot{Z}=\prod_{p,q}dx_{pq}\,dy_{pq}$, $x_{pq}+iy_{pq}=z_{pq}$. Then

$$J_{m,n}(\lambda)=\frac{\prod_{j=1}^{n}\Gamma(\lambda+j)\prod_{k=1}^{m}\Gamma(\lambda+k)}{\prod_{l=1}^{n+m}\Gamma(\lambda+l)}\pi^{mn}. \tag{2.2.1}$$

In particular, for $\lambda=0$ we get that the volume of the domain $\mathfrak{R}_{\mathrm{I}}$ of complex matrices Z such that $I-Z\bar{Z}'>0$ is equal to

$$V(\mathfrak{R}_{\mathrm{I}})=\frac{1!\,2!\ldots(m-1)!\,1!\,2!\ldots(n-1)!}{1!\,2!\ldots(m+n-1)!}\pi^{mn}. \tag{2.2.2}$$

First we deduce two recursive formulas for integrals of the type

$$\int_{I-Z\bar{Z}'>0} f(Z)\dot{Z}.$$

(1) We write the matrix Z in the form $Z=(Z_{m,n-1}, q)$, where $Z_{m,n-1}$ is an $m\times(n-1)$ matrix and q is a column. It is clear that

$$I-Z\bar{Z}'=I-Z_{m,\,n-1}\bar{Z}'_{m,\,n-1}-q\bar{q}'.$$

From $I-Z\bar{Z}'>0$ and $q\bar{q}'\geqq 0$ it follows that $I-Z_{m,\,n-1}\bar{Z}'_{m,\,n-1}>0$. Consequently, there exists a nonsingular matrix Γ such that

$$I-Z_{m,\,n-1}\bar{Z}'_{m,\,n-1}=\Gamma\bar{\Gamma}'.$$

We make the substitution $q=\Gamma w$. Then

$$\dot{q}=|\det\Gamma|^2\dot{w}=\dot{w}\det\Gamma\bar{\Gamma}'=\dot{w}\det\left(I-Z_{m,\,n-1}\bar{Z}'_{m,\,n-1}\right).$$

From this it follows that

$$\int_{I-Z\bar{Z}'>0} f(Z)\dot{Z}$$

$$=\int_{I-Z_{m,\,n-1}\bar{Z}'_{m,\,n-1}>0}\det\left(I-Z_{m,n-1}\bar{Z}'_{m,n-1}\right)\dot{Z}_{m,n-1}\int_{I-w\bar{w}'>0} f(Z)\dot{w}.$$

Here we used the formula

$$I-Z_{m,\,n-1}\bar{Z}'_{m,\,n-1}-q\bar{q}'=\Gamma(I-w\bar{w}')\bar{\Gamma}'.$$

Furthermore, by Theorem 2.1.2 the inequality $I-w\bar{w}'>0$ is equivalent to the inequality $1-\bar{w}'w>0$, which determines an ordinary hypersphere in the space. Therefore,

$$\int_{I-Z\bar{Z}'>0} f(Z)\dot{Z}$$

$$=\int_{I-Z_{m,\,n-1}\bar{Z}'_{m,\,n-1}>0}\det\left(I-Z_{m,\,n-1}\bar{Z}'_{m,\,n-1}\right)\dot{Z}_{m,\,n-1}\cdot\int_{1-\bar{w}'w>0} f(Z)\dot{w}. \tag{2.2.3}$$

(2) Now we represent Z in the form

$$Z=\begin{pmatrix} Z_{m-1,n}\\ p\end{pmatrix},$$

where $Z_{m-1,n}$ is an $(m-1)\times n$ matrix and p is a vector. Since the inequality $I-Z\bar{Z}'>0$ is equivalent to the inequality $I-\bar{Z}'Z>0$, analogously to (1) we obtain

$$\int_{I-Z\bar{Z}'>0} f(Z)\dot{Z} = \int_{I-\bar{Z}'Z>0} f(Z)\dot{Z}$$

$$= \int_{I-\bar{Z}'_{m-1,n}Z_{m-1,n}>0} \det\left(I-\bar{Z}'_{m-1,n}Z_{m-1,n}\right)\dot{Z}_{m-1,n} \int_{I-\bar{u}'u>0} f(Z)\dot{u},$$

where $p=u\Gamma$, and $I-\bar{Z}'_{m-1,n}Z_{m-1,n}=\Gamma\bar{\Gamma}'$. Hence, we find

$$\int_{I-Z\bar{Z}'>0} f(Z)\dot{Z}$$

$$= \int_{I-Z_{m-1,n}\bar{Z}'_{m-1,n}>0} \det\left(I-Z_{m-1,n}\bar{Z}'_{m-1,n}\right)\dot{Z}_{m-1,n}\cdot \int_{I-u\bar{u}'>0} f(Z)\dot{u}. \tag{2.2.4}$$

PROOF OF THEOREM 2.2.1. Applying (2.2.3) repeatedly we obtain

$$\int_{I-Z\bar{Z}'>0} f(Z)\dot{Z}$$

$$= \int_{w_1\bar{w}_1'<1} (1-w_1\bar{w}_1')^{n-1}\dot{w}_1 \int_{w_2\bar{w}_2'<1} (1-w_2\bar{w}_2')^{n-2}\dot{w}_2 \cdots \int_{w_n\bar{w}_n'<1} f(Z)\dot{w}_n. \tag{2.2.5}$$

Setting $f(Z)=\{\det(I-Z\bar{Z}')\}^\lambda$, we get

$$J_{m,n}(\lambda)=\prod_{j=1}^{n}\int_{w\bar{w}'<1}(1-w\bar{w}')^{j-1+\lambda}\dot{w}.$$

[This formula can also be deduced from (2.2.4).]

Using the well-known formula

$$\int\cdots\int_{x_1^2+\cdots+x_{2m}^2<1}(1-x_1^2-\cdots-x_{2m}^2)^{\mu-1}dx_1\ldots dx_{2m}=\pi^m\frac{\Gamma(\mu)}{\Gamma(\mu+m)} \quad (\mu>0), \tag{2.2.6}$$

we get the assertion of the theorem.

Using a similar method, based on a repeated application of (2.2.4), we can easily prove the following theorem.

THEOREM 2.2.2. *Let Z be an $m\times n$ matrix, $m>l$ and suppose that the function $f(Z)$ does not depend on the last $m-l$ rows of Z. Further, let Z_l be the matrix formed by the first l rows of Z, so that $f(Z)=f(Z_l)$. Then*

$$\int\limits_{I-Z\bar{Z}'>0} f(Z)\dot{Z}$$

$$=\pi^{n(m-l)}\frac{1!2!\dots(m-l-1)!}{n!(n+1)!\dots(n+m-l-1)!}\int\limits_{I-Z_l\bar{Z}_l'>0} f(Z_l)(\det(I-Z_l\bar{Z}_l'))^{m-l}\dot{Z}_l. \tag{2.2.7}$$

2.3. The volume of $\mathfrak{R}_{II}$.

THEOREM 2.3.1. *Let Z be a symmetric matrix of order n, and*

$$J_n(\lambda)=\int\limits_{I-Z\bar{Z}>0}\{\det(I-Z\bar{Z})\}^{\lambda}\dot{Z},$$

where

$$\dot{Z}=\prod_{p\leqslant q}dx_{pq}\,dy_{pq},\qquad x_{pq}+iy_{pq}=z_{pq}.$$

For $\lambda>-1$ we have

$$J_n(\lambda)=\frac{\pi^{\frac{n(n+1)}{2}}}{(\lambda+1)\dots(\lambda+n)}\cdot\frac{\Gamma(2\lambda+3)\Gamma(2\lambda+5)\dots\Gamma(2\lambda+2n-1)}{\Gamma(2\lambda+n+2)\Gamma(2\lambda+n+3)\dots\Gamma(2\lambda+2n)}. \tag{2.3.1}$$

In particular for $\lambda=0$ we get a formula for the volume of the domain $\mathfrak{R}_{II}$ of symmetric matrices Z satisfying the condition $I-Z\bar{Z}>0$,

$$V(\mathfrak{R}_{II})=J_n(0)=\frac{\pi^{\frac{n(n+1)}{2}}}{n!}\cdot\frac{2!4!\dots(2n-2)!}{(n+1)!(n+2)!\dots(2n-1)!}. \tag{2.3.2}$$

We need the following theorem.

THEOREM 2.3.2. *Let a and c be real numbers, b a complex number, connected by the inequalities $a<0$, $|b|^2-ac>0$, and let $\lambda>-1$. Then*

$$\iint\limits_{c+b\bar{z}+\bar{b}z+az\bar{z}>0}(c+b\bar{z}+\bar{b}z+az\bar{z})^{\lambda}\dot{z}=\frac{1}{|a|}\left(\frac{|b|^2-ac}{|a|}\right)^{\lambda+1}\cdot\frac{\pi}{\lambda+1}. \tag{2.3.3}$$

Proof. Making the substitution

$$w=\left(z+\frac{b}{a}\right)\sqrt{\frac{a^2}{|b|^2-ac}},$$

we obtain

$$\dot{w}=\frac{a^2}{|b|^2-ac}\dot{z},$$

$$c+b\bar{z}+\bar{b}z+az\bar{z}=c-\frac{|b|^2}{a}+a\left(z+\frac{b}{a}\right)\overline{\left(z+\frac{b}{a}\right)}$$

$$=\left(c-\frac{|b|^2}{a}\right)(1-w\bar{w}).$$

Then the left-hand side of (2.3.3) becomes

$$\frac{1}{|a|}\left(\frac{|b|^2-ac}{|a|}\right)^{\lambda+1}\iint_{1-w\bar{w}>0}(1-w\bar{w})^{\lambda}\dot{w}=\frac{1}{|a|}\left(\frac{|b|^2-ac}{|a|}\right)^{\lambda+1}\cdot\frac{\pi}{\lambda+1},$$

which we had to prove.

Proof of Theorem 2.3.1. We set

$$Z=\begin{pmatrix} Z_1 & v' \\ v & z\end{pmatrix},$$

where Z_1 is a symmetric matrix of order $n-1$, v is an $(n-1)$-dimensional vector, and z is a complex number. Then

$$I-Z\bar{Z}=\begin{pmatrix} I-Z_1\bar{Z}_1-v'\bar{v} & -(Z_1\bar{v}'+v'\bar{z}) \\ -(v\bar{Z}_1+z\bar{v}) & 1-v\bar{v}'-z\bar{z}\end{pmatrix}.$$

Using (2.1.9), we find that the condition $I-Z\bar{Z}>0$ is equivalent to the following two conditions.

$$I-Z_1\bar{Z}_1-v'\bar{v}>0,$$

$$1-v\bar{v}'-z\bar{z}-(v\bar{Z}_1+z\bar{v})(I-Z_1\bar{Z}_1-v'\bar{v})^{-1}\overline{(v\bar{Z}_1+z\bar{v})'}>0.$$

Besides,

$$\det(I-Z\bar{Z})=\det(I-Z_1\bar{Z}_1-v'\bar{v})$$

$$\times\{1-v\bar{v}'-z\bar{z}-(v\bar{Z}_1+z\bar{v})(I-Z_1\bar{Z}_1-v'\bar{v})^{-1}\overline{(v\bar{Z}_1+z\bar{v})'}\}.$$

Consequently,

$$J_n(\lambda)=\int_{I-Z_1\bar{Z}_1-v'\bar{v}>0}\{\det(I-Z_1\bar{Z}_1-v'\bar{v})\}^{\lambda}\dot{Z}_1\dot{v}$$

$$\times\iint_{c+b\bar{z}+\bar{b}z+az\bar{z}>0}(c+b\bar{z}+\bar{b}z+az\bar{z})^{\lambda}\dot{z},$$

where

$$a = -1 - \bar{v}(I - Z_1\bar{Z}_1 - v'\bar{v})^{-1} v' \qquad (a < 0),$$
$$b = -v\bar{Z}_1(I - Z_1\bar{Z}_1 - v'\bar{v})^{-1} v',$$
$$c = 1 - v\bar{v}' - v\bar{Z}_1(I - Z_1\bar{Z}_1 - v'\bar{v})^{-1} Z_1\bar{v}'.$$

Since the matrix $I - Z_1\bar{Z}_1$ is positive definite, there exists a nonsingular matrix Γ such that $I - Z_1\bar{Z}_1 = \Gamma\bar{\Gamma}'$. We make the substitution $v' = \Gamma u'$. Then $v = u\Gamma'$, and

$$\dot{v} = |\det \Gamma|^2 \dot{u} = \dot{u} \det(I - Z_1\bar{Z}_1),$$
$$(I - Z_1\bar{Z}_1 - v'\bar{v})^{-1} = \bar{\Gamma}'^{-1}(I - u'\bar{u})^{-1}\Gamma^{-1}.$$

Furthermore, since

$$(I - u'\bar{u})^{-1}u' = \frac{u'}{1 + \bar{u}u'}, \qquad w(I - u'\bar{u})^{-1}\bar{w}' = w\bar{w}' + \frac{|wu'|^2}{1 - \bar{u}u'},$$

(w is an arbitrary $(n-1)$-dimensional vector), we obtain

$$a = -\left(1 + \bar{u}(I - u'\bar{u})^{-1} u'\right) = -\frac{1}{1 - \bar{u}u'},$$
$$b = -u\Gamma'\bar{Z}_1\bar{\Gamma}'^{-1}(I - u'\bar{u})^{-1} u' = -\frac{u\Gamma'\bar{Z}_1\bar{\Gamma}'^{-1}u'}{1 - \bar{u}u'},$$
$$c = 1 - u\Gamma'\bar{\Gamma}\bar{u}' - u\Gamma'\bar{Z}_1\bar{\Gamma}'^{-1}(I - u'\bar{u})^{-1}\Gamma^{-1}Z_1\bar{\Gamma}\bar{u}'$$
$$= 1 - u\Gamma'\bar{\Gamma}\bar{u}' - u\Gamma'\bar{Z}_1\bar{\Gamma}'^{-1}\Gamma^{-1}Z_1\bar{\Gamma}\bar{u}' - \frac{|u\Gamma'\bar{Z}_1\bar{\Gamma}'^{-1}u'|^2}{1 - \bar{u}u'}.$$

Consequently,

$$|b|^2 - ac = \frac{1}{1 - \bar{u}u'}\left(1 - u\Gamma'\bar{\Gamma}\bar{u}' - u\Gamma'\bar{Z}_1\bar{\Gamma}'^{-1}\Gamma^{-1}Z_1\bar{\Gamma}\bar{u}'\right)$$
$$= \frac{1}{1 - \bar{u}u'}\left\{1 - u\Gamma'\left(I + \bar{Z}_1(I - Z_1\bar{Z}_1)^{-1}Z_1\right)\bar{\Gamma}\bar{u}'\right\}$$
$$= \frac{1}{1 - \bar{u}u'}\left\{1 - u\Gamma'(I - \bar{Z}_1Z_1)^{-1}\bar{\Gamma}\bar{u}'\right\} = 1.$$

Then, by Theorem 2.3.2 and using the formula

$$\det(I - Z_1\bar{Z}_1 - v'\bar{v}) = \det(1 - u'\bar{u})\det(I - Z_1\bar{Z}_1)$$
$$= (1 - \bar{u}u')\det(I - Z_1\bar{Z}_1),$$

we obtain

$$J_n(\lambda) = \frac{\pi}{\lambda+1} \int\limits_{I - Z_1\bar{Z}_1 - v'\bar{v} > 0} \frac{\{\det(I - Z_1\bar{Z}_1 - v'\bar{v})\}^{\lambda}}{\{1 + \bar{v}(I - Z_1\bar{Z}_1 - v'\bar{v})^{-1} v'\}^{\lambda+2}} \dot{Z}_1 \dot{v}$$

$$= \frac{\pi}{\lambda+1} \int\limits_{I - Z_1\bar{Z}_1 > 0} \{\det(I - Z_1\bar{Z}_1)\}^{\lambda+1} \dot{Z}_1 \cdot \int\limits_{1 - \bar{u}u' > 0} (1 - \bar{u}u')^{2\lambda+2} \dot{u}$$

(using Theorem 2.2.2). By (2.2.6) we have

$$J_n(\lambda) = J_{n-1}(\lambda+1) \cdot \frac{\pi^n}{\lambda+1} \cdot \frac{\Gamma(2\lambda+3)}{\Gamma(2\lambda+n+2)}. \tag{2.3.4}$$

Continuing this reduction down to $n=1$, we get

$$J_n(\lambda) = \frac{\pi^{\frac{n(n+1)}{2}}}{(\lambda+1)\dots(\lambda+n)} \cdot \frac{\Gamma(2\lambda+3)\Gamma(2\lambda+5)\dots\Gamma(2\lambda+2n-1)}{\Gamma(2\lambda+n+2)\Gamma(2\lambda+n+3)\dots\Gamma(2\lambda+2n)}.$$

2.4. The volume of $\mathfrak{R}_{\mathrm{III}}$.

THEOREM 2.4.1. *Let $n \geqq 2$ and let Z be a skew-symmetric matrix of order n. We set*

$$K_n(\lambda) = \int\limits_{I + Z\bar{Z} > 0} \{\det(I + Z\bar{Z})\}^{\lambda} \dot{Z},$$

where

$$\dot{Z} = \prod_{p<q} dx_{pq}\, dy_{pq}, \quad x_{pq} + iy_{pq} = z_{pq}.$$

For $\lambda > -\frac{1}{2}$,

$$K_n(\lambda) = \pi^{\frac{n(n-1)}{2}} \frac{\Gamma(2\lambda+1)\Gamma(2\lambda+3)\dots\Gamma(2\lambda+2n-3)}{\Gamma(2\lambda+n)\Gamma(2\lambda+n+1)\dots\Gamma(2\lambda+2n-2)}. \tag{2.4.1}$$

In particular, for $\lambda=0$ we obtain a formula for the volume of the domain $\mathfrak{R}_{\mathrm{III}}$ of skew-symmetric matrices Z, satisfying the condition $I + Z\bar{Z} > 0$,

$$V(\mathfrak{R}_{\mathrm{III}}) = \pi^{\frac{n(n-1)}{2}} \frac{2!\,4!\dots(2n-4)!}{(n-1)!\,n!\dots(2n-3)!}. \tag{2.4.2}$$

PROOF. We set

$$Z = \begin{pmatrix} Z_1 & -u' \\ u & 0 \end{pmatrix},$$

where Z_1 is a skew-symmetric matrix of order $n-1$, and u is an $(n-1)$-dimensional vector. Then

$$I+Z\bar{Z}=\begin{pmatrix} I+Z_1\bar{Z}_1-u'\bar{u} & -Z_1\bar{u}' \\ u\bar{Z}_1 & 1-u\bar{u}' \end{pmatrix}.$$

By (2.1.9) the condition $I+Z\bar{Z}>0$ is equivalent to the following two conditions

$$I+Z_1\bar{Z}_1-u'\bar{u}>0, \tag{2.4.3}$$

$$1-u\bar{u}'+u\bar{Z}_1(I+Z_1\bar{Z}_1-u'\bar{u})^{-1}Z_1\bar{u}'>0. \tag{2.4.4}$$

Furthermore,

$$\det(I+Z\bar{Z})$$
$$=\left(1-u\bar{u}'+u\bar{Z}_1(I+Z_1\bar{Z}_1-u'\bar{u})^{-1}Z_1\bar{u}'\right)\det(I+Z_1\bar{Z}_1-u'\bar{u}).$$

Let the matrix Γ again satisfy the condition $I+Z_1\bar{Z}_1=\Gamma\bar{\Gamma}$. We set $u=v\Gamma'$. By (2.4.3) we have $I+Z_1\bar{Z}_1>0$ and $I-v'\bar{v}>0$. The left-hand side of the inequality (2.4.4) transforms into

$$1-v\Gamma'\bar{\Gamma}\bar{v}'+v\Gamma'\bar{Z}_1\bar{\Gamma}'^{-1}(I-v'\bar{v})^{-1}\Gamma^{-1}Z_1\bar{\Gamma}\bar{v}'$$
$$=1-v\Gamma'\bar{\Gamma}\bar{v}'+v\Gamma'\bar{Z}_1\bar{\Gamma}'^{-1}\Gamma^{-1}Z_1\bar{\Gamma}\bar{v}'-\frac{|v\Gamma'\bar{Z}_1\bar{\Gamma}'^{-1}v'|^2}{1-\bar{v}v'}$$
$$=1-v\bar{v}'-\frac{|v\Gamma'\bar{Z}_1\bar{\Gamma}'^{-1}v'|^2}{1-\bar{v}v'}=1-v\bar{v}'.$$

(Here we have used the fact that the matrix $\Gamma'\bar{Z}_1\bar{\Gamma}'^{-1}$ is skew-symmetric.) Consequently,

$$K_n(\lambda)=\int_{I+Z_1\bar{Z}_1>0}\{\det(I+Z_1\bar{Z}_1)\}^{\lambda+1}\dot{Z}_1\cdot\int_{1-v'\bar{v}>0}(1-v\bar{v}')^{2\lambda}\dot{v}$$
$$=\pi^{n-1}\frac{\Gamma(2\lambda+1)}{\Gamma(2\lambda+n)}K_{n-1}(\lambda+1).$$

Continuing the reduction, we obtain the statement of the theorem, since

$$K_2(\lambda+n-2)=\iint_{|z|<1}(1-|z|^2)^{2\lambda+2n-4}\dot{z}=\pi\frac{\Gamma(2\lambda+2n-3)}{\Gamma(2\lambda+2n-2)}.$$

2.5. **The volume of** $\mathfrak{R}_{\mathrm{IV}}$. The domain $\mathfrak{R}_{\mathrm{IV}}$ (the Lie-sphere) consists of all n-dimensional complex vectors z satisfying the conditions

$$|zz'|^2 + 1 - 2\bar{z}z' > 0, \tag{2.5.1}$$

$$|zz'| < 1. \tag{2.5.2}$$

We show that these two inequalities can be replaced by one. First we re-write the inequalities (2.5.1) and (2.5.2) as

$$(1 - \bar{z}z')^2 > (\bar{z}z')^2 - |zz'|^2 > (\bar{z}z')^2 - 1. \tag{2.5.3}$$

From (2.5.3) it follows easily that

$$\bar{z}z' < 1. \tag{2.5.4}$$

The first of the inequalities (2.5.3) implies (since $\bar{z}z' \geqq |zz'|$) that

$$1 - \bar{z}z' > \sqrt{(\bar{z}z')^2 - |zz'|^2}. \tag{2.5.5}$$

Therefore all vectors z satisfying the inequalities (2.5.1) and (2.5.2) also satisfy the inequality (2.5.5).

On the other hand, every vector satisfying (2.5.5) obviously satisfies (2.5.1). Furthermore from $|zz'| \leqq \bar{z}z'$ and (2.5.5) we can deduce $|zz'| < 1$, i.e., (2.5.2). Therefore the domain $\mathfrak{R}_{\mathrm{IV}}$ can be defined by the single inequality (2.5.5).

THEOREM 2.5.1. *For* $\alpha > -1$ *and* $\beta > -(n+\alpha)$

$$L_n(\alpha, \beta) = \int_{\mathfrak{R}_{\mathrm{IV}}} \left(1 - \bar{z}z' - \sqrt{(\bar{z}z')^2 - |zz'|^2}\right)^{\alpha}$$

$$\times \left(1 - \bar{z}z' + \sqrt{(\bar{z}z')^2 - |zz'|^2}\right)^{\beta} \dot{z} = \frac{\pi^n}{2^{n-1}} \frac{\Gamma(\alpha+1)}{\Gamma(\alpha+n)} \cdot \frac{1}{\alpha+\beta+n}. \tag{2.5.6}$$

In particular, for $\alpha = \beta = 0$ *we obtain a formula for the volume of the domain* $\mathfrak{R}_{\mathrm{IV}}$,

$$V(\mathfrak{R}_{\mathrm{IV}}) = \frac{\pi^n}{2^{n-1}} \cdot \frac{1}{n!}. \tag{2.5.7}$$

PROOF. For $n = 1$ we set $z = x + iy$, where x and y are real numbers. Then

$$\bar{z}z' = |zz'| = x^2 + y^2,$$

and therefore, $\mathfrak{R}_{IV}$ is the unit disc in the complex plane. Consequently

$$L_1(\alpha, \beta) = \iint\limits_{x^2+y^2<1} (1 - x^2 - y^2)^{\alpha+\beta}\, dx\, dy = \frac{\pi}{\alpha+\beta+1}.$$

So for $n=1$ the theorem is proved.

For $n \geqq 2$ we set $z = x + iy$, where x and y are real vectors. The inequality (2.5.5) has the form

$$1 - xx' - yy' > 2\sqrt{xx'yy' - (xy')^2}. \tag{2.5.8}$$

Hence

$$L_n(\alpha, \beta) = \int\limits_{x,\,y} \left(1 - xx' - yy' - 2\sqrt{xx'yy' - (xy')^2}\right)^{\alpha}$$
$$\times \left(1 - xx' - yy' + 2\sqrt{xx'yy' - (xy')^2}\right)^{\beta} \dot{x}\dot{y},$$

where the integral is taken on the domain described by the inequality (2.5.8). For every fixed x we can find an orthogonal matrix R with determinant 1 such that

$$xR = (\sqrt{xx'}, 0, \ldots, 0).$$

We set $yR = (\xi, w)$ where ξ is some real number, and w is an $(n-1)$-dimensional vector. Then (2.5.8) transforms into

$$1 - xx' - \xi^2 - ww' > 2\sqrt{xx'(\xi^2 + ww') - xx'\xi^2} = 2\sqrt{xx'ww'}. \tag{2.5.9}$$

Consequently, after the substitution, the integral transforms into

$$L_n(\alpha, \beta) = \int\limits_{\xi,\,w,\,x} \left(1 - \xi^2 - xx' - ww' - 2\sqrt{xx'ww'}\right)^{\alpha}$$
$$\times \left(1 - \xi^2 - xx' - ww' + 2\sqrt{xx'ww'}\right)^{\beta} d\xi\, \dot{w}\dot{x},$$

where the integral is taken on the domain described by the inequality (2.5.9). Furthermore, making the substitution

$$x = u\sqrt{(1-\xi^2)}, \qquad w = v\sqrt{(1-\xi^2)},$$

we obtain

$$L_n(\alpha, \beta) = \int_{-1}^{1} (1-\xi^2)^{\alpha+\beta+n-\frac{1}{2}} d\xi \int\limits_{1-uu'-vv' > 2\sqrt{uu'vv'}} \left(1 - uu' - vv' - 2\sqrt{uu'vv'}\right)^{\alpha}$$

$$\times \left(1 - uu' - vv' + 2\sqrt{uu'vv'}\right)^{\beta} \dot{u}\dot{v} = \pi^{\frac{1}{2}} \frac{\Gamma\left(\alpha+\beta+n+\frac{1}{2}\right)}{\Gamma(\alpha+\beta+n+1)} 2^{2n-1} \cdot P, \tag{2.5.10}$$

where

$$P = \int\limits_{\substack{1-uu'-vv' > 2\sqrt{uu'vv'} \\ u_\mu \geqslant 0,\ v_\nu \geqslant 0}} \left(1 - uu' - vv' - 2\sqrt{uu'vv'}\right)^{\alpha} \times \left(1 - uu' - vv' + 2\sqrt{uu'vv'}\right)^{\beta} \dot{u}\dot{v}.$$

Now we set $\eta^2 = uu'$, $\zeta^2 = vv'$. Then we obtain

$$P = \iint\limits_{\substack{\eta+\zeta<1 \\ \eta \geqslant 0,\ \zeta \geqslant 0}} (1-(\eta+\zeta)^2)^{\alpha} (1-(\eta-\zeta)^2)^{\beta} \, d\eta \, d\zeta \times \int \cdots \int\limits_{\substack{u_2^2+\cdots+u_n^2 \leqslant \eta^2 \\ u_\mu \geqslant 0}} \frac{\eta \, du_2 \ldots du_n}{\sqrt{\eta^2 - u_2^2 - \cdots - u_n^2}} \times \int \cdots \int\limits_{\substack{v_3^2+\cdots+v_n^2 \leqslant \zeta^2 \\ v_\nu \geqslant 0}} \frac{\zeta \, dv_3 \ldots dv_n}{\sqrt{\zeta^2 - v_3^2 - \cdots - v_n^2}}. \tag{2.5.11}$$

Using that

$$\int \cdots \int\limits_{\substack{u_2^2+\cdots+u_n^2 \leqslant \eta^2 \\ u_\mu \geqslant 0}} \frac{\eta \, du_2 \ldots du_n}{\sqrt{\eta^2 - u_2^2 - \cdots - u_n^2}}$$

$$= \eta^{n-1} \int \cdots \int\limits_{\substack{u_2^2+\cdots+u_n^2 \leqslant 1 \\ u_\mu \geqslant 0}} \frac{du_2 \ldots du_n}{\sqrt{1 - u_2^2 - \cdots - u_n^2}} = \eta^{n-1} \frac{\pi^{\frac{n}{2}}}{\Gamma\left(\frac{n}{2}\right)} \frac{1}{2^{n-1}}, \tag{2.5.12}$$

the formula (2.5.11) gives

$$P=\frac{\pi^{n-\frac{1}{2}}2^{-(2n-3)}}{\Gamma\left(\frac{n}{2}\right)\Gamma\left(\frac{n-1}{2}\right)}\underset{\substack{\eta+\zeta\leqslant 1\\ \eta\geqslant 0,\,\zeta\geqslant 0}}{\int\int}(1-(\eta+\zeta)^2)^\alpha(1-(\eta-\zeta)^2)^\beta\,\eta^{n-1}\zeta^{n-2}\,d\eta\,d\zeta$$

$$=\frac{\pi^{n-\frac{1}{2}}2^{-(2n-3)}}{\Gamma\left(\frac{n}{2}\right)\Gamma\left(\frac{n-1}{2}\right)}\underset{\substack{\eta+\zeta\leqslant 1\\ 0\leqslant\eta\leqslant\zeta}}{\int\int}(1-(\eta+\zeta)^2)^\alpha$$

$$\times(1-(\eta-\zeta)^2)^\beta\eta^{n-2}\zeta^{n-2}(\eta+\zeta)\,d\eta\,d\zeta$$

or, setting $\zeta-\eta=\tau$, $\zeta+\eta=\sigma$,

$$P=\frac{\pi^{n-\frac{1}{2}}2^{-(2n-3)}}{\Gamma\left(\frac{n}{2}\right)\Gamma\left(\frac{n-1}{2}\right)}\underset{0\leqslant\tau\leqslant\sigma\leqslant 1}{\int\int}(1-\sigma^2)^\alpha(1-\tau^2)^\beta\left(\frac{\sigma+\tau}{2}\right)^{n-2}\left(\frac{\sigma-\tau}{2}\right)^{n-2}\sigma\,\frac{d\sigma\,d\tau}{2}$$

$$=\frac{\pi^{n-\frac{1}{2}}2^{-(4n-6)}}{\Gamma\left(\frac{n}{2}\right)\Gamma\left(\frac{n-1}{2}\right)}\int_0^1(1-\tau^2)^\beta\,d\tau\int_\tau^1(1-\sigma^2)^\alpha(\sigma^2-\tau^2)^{n-2}\sigma\,d\sigma.$$

Making the substitution $\omega=(\sigma^2-\tau^2)/(1-\tau^2)$, $1-\omega=(1-\sigma^2)/(1-\tau^2)$ in the inside integral, we have

$$P=\frac{\pi^{n-\frac{1}{2}}2^{-(4n-5)}}{\Gamma\left(\frac{n}{2}\right)\Gamma\left(\frac{n-1}{2}\right)}\int_0^1(1-\tau^2)^{\alpha+\beta+n-1}d\tau\int_0^1(1-\omega)^\alpha\omega^{n-2}\,d\omega$$

$$=\frac{\pi^{n-\frac{1}{2}}2^{-(4n-4)}}{\Gamma\left(\frac{n}{2}\right)\Gamma\left(\frac{n-1}{2}\right)}\cdot\frac{\pi^{\frac{1}{2}}\Gamma(\alpha+\beta+n)}{\Gamma\left(\alpha+\beta+n+\frac{1}{2}\right)}\cdot\frac{\Gamma(\alpha+1)\Gamma(n-1)}{\Gamma(\alpha+n)}.$$

Substituting this formula into (2.5.10) we obtain

$$L_n(\alpha,\beta)=\frac{1}{2^{2n-3}}\,\frac{\pi^{n+\frac{1}{2}}\Gamma(n-1)}{\Gamma\left(\frac{n}{2}\right)\Gamma\left(\frac{n-1}{2}\right)}\cdot\frac{\Gamma(\alpha+\beta+n)}{\Gamma(n+\alpha+\beta+1)}\cdot\frac{\Gamma(\alpha+1)}{\Gamma(\alpha+n)}$$

$$=\frac{\pi^n}{2^{n-1}}\,\frac{\Gamma(\alpha+1)}{\Gamma(\alpha+n)}\cdot\frac{1}{\alpha+\beta+n}.$$

(We have used the formula $\Gamma(x)\Gamma(x+\frac{1}{2})=\pi^{1/2}\Gamma(2x)2^{1-2x}$.)

CHAPTER III

POLAR COORDINATES FOR MATRICES

3.1. **The volume element of the space of unitary matrices.** If the square matrix Z of order n is regarded as a point in $2n^2$-dimensional real space, then the set of all unitary matrices is an n^2-dimensional manifold in this space. This manifold, as well as the group of unitary matrices itself, will be denoted by U_n. Now we are going to find the volume element of this manifold.

If the $2n^2$-dimensional space of matrices of Z is regarded as a usual Euclidean space, then the first quadratic form of this space can be written as

$$\mathrm{Sp}\,(dZ\,d\bar{Z}') = \sum_{i,\,j=1}^{n} |\,dz_{ij}\,|^2.$$

If $Z = U$ (a unitary matrix), then since

$$d\bar{U}' = dU^{-1} = -U^{-1}\cdot dU \cdot U^{-1},$$

we get the following expression for the first quadratic form of the manifold $\mathfrak{U}_n$

$$ds^2 = -\,\mathrm{Sp}\,(dU\cdot U^{-1}\cdot dU\cdot U^{-1}). \tag{3.1.1}$$

We set

$$\delta U = U^{-1}\,dU = \bar{U}'\,dU. \tag{3.1.2}$$

Since

$$0 = dI = d\,(\bar{U}'U) = \bar{U}'\,dU + d\bar{U}'\cdot U,$$

we have

$$\overline{\delta U}' = -\,\delta U. \tag{3.1.3}$$

Consequently, (3.1.1) can be rewritten in the form

$$ds^2 = \mathrm{Sp}\,(\delta U\cdot \overline{\delta U}'). \tag{3.1.4}$$

Writing δU in the form (δu_{jk}), from (3.1.3) we find $\delta u_{jk} = -\overline{\delta u_{jk}}$. Substituting into (3.1.4) we obtain

$$ds^2 = \sum_{j,k=1}^{n} |\delta u_{jk}|^2. \tag{3.1.5}$$

Introducing the real quantities

$$\delta u_{jj} = i\delta s_j, \quad \delta u_{jk} = \delta s_{jk} + i\delta s'_{jk}, \quad \delta u_{kj} = -\delta s_{jk} + i\delta s'_{jk}, \quad j < k,$$

we have

$$ds^2 = \sum_{j=1}^{n} \delta s_j^2 + 2\sum_{j<k} \left(\delta s_{jk}^2 + \delta s_{jk}'^2\right). \tag{3.1.6}$$

That is, the volume element of $\mathfrak{U}_n$ can be represented in the form

$$\dot{U} = 2^{\frac{n(n-1)}{2}} \prod_{j=1}^{n} \delta s_j \prod_{j<k} \delta s_{jk} \delta s'_{jk}. \tag{3.1.7}$$

Now, if V and W are two fixed unitary matrices, then performing the transformation

$$U_1 = VUW, \tag{3.1.8}$$

we find

$$\delta U_1 = \overline{W}' \delta U W,$$

whence

$$\mathrm{Sp}\left(\delta U_1 \cdot \overline{\delta U_1'}\right) = \mathrm{Sp}\,(\overline{W}' \cdot \delta U \cdot \overline{\delta U}' \cdot W) = \mathrm{Sp}\,(\delta U \cdot \overline{\delta U'}). \tag{3.1.9}$$

So we have established that the volume element of the space of unitaries is invariant under transformations of the form (3.1.8). This fact can be summarized in the form

$$\dot{U}_1 = \dot{U}. \tag{3.1.10}$$

Now we introduce a parametric representation for unitary matrices. We set

$$U = (I + iH)(I - iH)^{-1}. \tag{3.1.11}$$

It is clear that the matrix H must be hermitian. We shall call this matrix H the parameter. This representation will often make the computation of integrals over the unitary matrices more intuitive. The inverse formula of (3.1.11) is

$$H = i(I - U)(I + U)^{-1}. \tag{3.1.12}$$

It is easy to show that this correspondence between unitary and hermitian matrices is one-to-one for almost all matrices. By "almost all" we mean that there exists an exceptional manifold of lower dimension, namely, the one for which $\det(I+U)=0$ or $\det(I-iH)=0$.

Differentiating (3.1.11) we obtain

$$dU = i\,dH\,(I-iH)^{-1} + i\,(I+iH)(I-iH)^{-1}\,dH\,(I-iH)^{-1}$$
$$= 2i\,(I-iH)^{-1}\,dH\,(I-iH)^{-1}.$$

That is

$$\delta U = 2i\,(I+iH)^{-1}\,dH\,(I-iH)^{-1}. \tag{3.1.13}$$

Hence we find

$$\mathrm{Sp}\,(\delta U\,\overline{\delta U'}) = 4\,\mathrm{Sp}\,\{(I+H^2)^{-1}\,dH\,(I+H^2)^{-1}\,dH\}.$$

Therefore,

$$\dot{U} = 2^{n^2}\{\det(I+H^2)\}^{-n}\dot{H}, \tag{3.1.14}$$

where

$$\dot{H} = 2^{\frac{n(n-1)}{2}}\prod_{j=1}^{n} dh_j \prod_{j<k} dh'_{jk}\,dh''_{jk},$$
$$H=(h_{jk}),\quad h_{jj}=h_j,\quad h_{jk}=h'_{jk}+ih''_{jk}.$$

In this representation the manifold of unitary matrices is determined by the inequalities

$$-\infty < h_j < \infty,\quad -\infty < h'_{jk},\ h''_{jk} < \infty, \tag{3.1.15}$$

with the exception of a manifold of lower dimension.

THEOREM 3.1.1. *The volume of the manifold* $\mathfrak{U}_n$ *of unitary matrices is equal to*

$$\omega_n = \frac{(2\pi)^{\frac{n(n+1)}{2}}}{1!\,2!\,\ldots\,(n-1)!}.$$

PROOF. From (3.1.4) we obtain

$$\omega_n = \int_U \dot{U} = 2^{n^2}\int_{-\infty}^{\infty}\cdots\int_{-\infty}^{\infty}\{\det(I+H^2)\}^{-n}\dot{H}.$$

However, from Theorem 2.1.5 we know that this integral equals

$$2^{\frac{n(n-1)}{2}} \pi^{\frac{n^2}{2}} \frac{\prod_{j=0}^{n-1} \Gamma\left(n-j-\frac{1}{2}\right) \prod_{k=0}^{n-2} \Gamma(n-k)}{\prod_{j=0}^{n-1} \Gamma(n-j) \prod_{k=0}^{n-2} \Gamma(2n-2k-1)}.$$

Since $\Gamma(x)\Gamma(x+\frac{1}{2})=\sqrt{\pi}\,\Gamma(2x)2^{1-2x}$, we have

$$\prod_{j=0}^{n-1} \Gamma(n-j)\Gamma\left(n-j-\frac{1}{2}\right) = \pi^{\frac{n}{2}} \prod_{j=0}^{n-1} \Gamma(2n-2j-1)\, 2^{2-2n+2j}$$

$$= \pi^{\frac{n}{2}} 2^{-n(n-1)} \prod_{k=0}^{n-2} \Gamma(2n-2k-1),$$

which completes the proof of the theorem.

REMARK 1. In some books the volume element of the unitary group is expressed as

$$\prod_{j=1}^{n} \delta s_j \prod_{j<k} \delta s_{jk}\, \delta s'_{jk},$$

which can be obtained from (3.1.7) by multiplying by $2^{n(n-1)/2}$.

REMARK 2. The exceptional case in the parametrical presentation (3.1.11) can be removed by introducing homogeneous coordinates for the hermitian matrices, i.e., using the methods of projective geometry, but we shall not go into the details of this.

3.2. **Integrals on a coset space of the unitary group.** Any unitary matrix can be represented in the form

$$U = V\Lambda V^{-1}, \quad \Lambda = [e^{i\theta_1}, \ldots, e^{i\theta_n}], \quad 2\pi > \theta_1 \geqslant \theta_2 \geqslant \cdots \geqslant \theta_n \geqslant 0, \quad (3.2.1)$$

where V is a unitary matrix. Since $e^{i\theta_1}, \cdots, e^{i\theta_n}$ are the eigenvalues of U, The matrix Λ is uniquely determined. If, furthermore,

$$U = V\Lambda V^{-1} = V_1 \Lambda V_1^{-1}$$

and if the eigenvalues of U are distinct, then

$$V_1^{-1} V = [e^{i\varphi_1}, \ldots, e^{i\varphi_n}].$$

The diagonal unitary matrices form a subgroup of the group $\mathfrak{U}_n$. We denote by $[\mathfrak{U}_n]$ the manifold of left cosets of $\mathfrak{U}_n$ with respect to this subgroup. Clearly, the above matrices V and V_1 belong to the same coset, i.e., the same element of $[\mathfrak{U}_n]$ corresponds to both of them. Thus, to almost every matrix U there corresponds a unique diagonal matrix Λ and an element from $[\mathfrak{U}_n]$. We shall clarify the relations between the volume elements of $\mathfrak{U}_n$, $[\mathfrak{U}_n]$ and the manifold of diagonal unitary matrices Λ.

Differentiating (3.2.1) we find

$$dU = dV \cdot \Lambda V^{-1} - V\Lambda V^{-1} \cdot dV \cdot V^{-1} + V \cdot d\Lambda \cdot V^{-1},$$

whence

$$\overline{V}' \cdot dU \cdot V = \delta V \cdot \Lambda - \Lambda\, \delta V + d\Lambda.$$

That is

$$\begin{aligned}\mathrm{Sp}\,(dU \cdot \overline{dU}') &= \mathrm{Sp}\,\{(\delta V \cdot \Lambda - \Lambda\,\delta V)(\overline{\delta V \cdot \Lambda - \Lambda\,\delta V})'\}\\ &+ \mathrm{Sp}\,\{(\delta V \cdot \Lambda - \Lambda \cdot \delta V)\,\overline{d\Lambda}'\} + \mathrm{Sp}\,\{d\Lambda \cdot (\overline{\delta V \cdot \Lambda - \Lambda \cdot \delta V})'\} + \mathrm{Sp}\,(d\Lambda \cdot \overline{d\Lambda}').\end{aligned}$$

But $\mathrm{Sp}\{(\delta V \cdot \Lambda - \Lambda \cdot \delta V)\overline{d\Lambda}'\} = 0$ and therefore, setting $\delta V = (\delta v_{jk})$, we have

$$\mathrm{Sp}\,(dU \cdot \overline{dU}') = \sum_{j,\,k=1}^{n} |\, \delta v_{jk} (e^{i\theta_j} - e^{i\theta_k})\,|^2 + \sum_{j=1}^{n} d\theta_j^2.$$

We note that the δv_{jj} in this formula disappear. We denote by $\dot{V}$ the volume element formed by the vectors δv_{jk} $(j \neq k)$. The volume element of $[\mathfrak{U}_n]$ is equal to $[\dot{U}] = 2^{n(n-1)/2}\,\dot{V}$. Thus

$$\dot{U} = [\dot{U}] \cdot \prod_{j<k} |\, e^{i\theta_j} - e^{i\theta_k} \,|^2\, d\theta_1 \ldots d\theta_n. \qquad (3.2.2)$$

THEOREM 3.2.1. *The volume of the coset manifold $[\mathfrak{U}_n]$ is equal to*

$$\omega_n' = \frac{(2\pi)^{\frac{n(n-1)}{2}}}{1!\,2! \ldots (n-1)!}.$$

PROOF. We have

$$\begin{aligned}\omega_n &= \omega_n' \int_0^{2\pi} d\theta_1 \int_0^{\theta_1} d\theta_2 \ldots \int_0^{\theta_{n-1}} \prod_{j<k} |\, e^{i\theta_j} - e^{i\theta_k} \,|^2\, d\theta_n\\ &= \frac{\omega_n'}{n!} \int_0^{2\pi} \cdots \int_0^{2\pi} \prod_{j<k} |\, e^{i\theta_j} - e^{i\theta_k} \,|^2\, d\theta_1 \ldots d\theta_n.\end{aligned}$$

Writing $\prod_{j<k} (e^{i\theta_j} - e^{i\theta_k})$ in the form

$$\sum_{j_1, \ldots, j_n} \delta_{j_1 \ldots j_n}^{1 \ldots n}\, e^{i\{(n-1)\theta_{j_1} + (n-2)\theta_{j_2} + \ldots + \theta_{j_{n-1}}\}}$$

($n!$ terms) and using the orthogonality of $e^{i(m_1\theta_1+\cdots+m_n\theta_n)}$ we have

$$\omega_n=(2\pi)^n\,\omega'_n,$$

whence our assertion follows.

Now we consider the set $\{\mathfrak{U}_n\}$ of cosets of the group $\mathfrak{U}_n$ with respect to its subgroup consisting of diagonal matrices of the form $[\pm 1, \pm 1, \cdots, \pm 1]$. Any unitary matrix U can be described by an element of $\{\mathfrak{U}_n\}$ and a diagonal matrix $\Lambda=[e^{i\theta_1}, \cdots, e^{i\theta_n}]$ $(0\leqq\theta_1\leqq\theta_2\leqq\cdots\leqq\theta_n<2\pi)$. Furthermore, any element of $\{\mathfrak{U}_n\}$ can be described by an element in $[\mathfrak{U}_n]$ and a diagonal matrix

$$\Lambda_1=[e^{i\theta_1}, \ldots, e^{i\theta_n}], \qquad 0\leqslant\theta_1\leqslant\ \ldots\ \leqslant\theta_n<\pi.$$

It is clear that the volume of $\{\mathfrak{U}_n\}$ is equal to $2^{-n}\omega_n$.

3.3. **Polar coordinates for hermitian matrices.** It is known that every hermitian matrix can be represented in the form

$$H=U\Lambda\bar{U}', \tag{3.3.1}$$

where U is a unitary matrix, and $\Lambda=[\lambda_1, \cdots, \lambda_n]$ is a diagonal matrix with $\lambda_1\geqq\lambda_2\geqq\cdots\geqq\lambda_n$. The pair of matrices (U, Λ) can be called the polar coordinates of the hermitian matrix. The matrix Λ corresponds to the absolute value in ordinary polar coordinates, and U to the argument. However, this correspondence is not one-to-one. In fact, from the equalities

$$H=U\Lambda\bar{U}'=U_1\Lambda_1\bar{U}'_1$$

it follows only that $\Lambda=\Lambda_1$ and $U_1^{-1}U=[e^{i\varphi_1}, \cdots, e^{i\varphi_n}]$ (in case the eigenvalues of H are all different). Therefore, a one-to-one correspondence will hold between H and $([\mathfrak{U}_n], \Lambda)$ (with the exception of the matrices H having multiple eigenvalues, i.e., with the exception of a manifold of lower dimension).

Differentiating (3.3.1), we obtain

$$dH=dU\cdot\Lambda\bar{U}'+U\,d\Lambda\cdot\bar{U}'+U\Lambda\,d\bar{U}',$$

whence

$$\bar{U}'\,dH\cdot U=\delta U\cdot\Lambda+d\Lambda-\Lambda\,\delta U.$$

Setting $\overline{U}'\cdot dH\cdot U=\delta G$ and $\delta G=(\delta g_{jk})$, we have

$$\delta g_{jj}=d\lambda_j, \quad \delta g_{jk}=\delta u_{jk}(\lambda_k-\lambda_j) \qquad (j\neq k).$$

Separating the real and imaginary parts, we obtain

$$\dot{H} = \prod_{j<k} (\lambda_j - \lambda_k)^2 [\dot{U}]\, d\lambda_1 \ldots d\lambda_n. \tag{3.3.2}$$

The last equality yields

$$\{\det(I + H^2)\}^{-n} \dot{H} = \prod_{j<k} (\lambda_j - \lambda_k)^2 \prod_{m=1}^{n} \left(1 + \lambda_m^2\right)^{-n} [\dot{U}]\, d\lambda_1 \ldots d\lambda_n. \tag{3.3.3}$$

Since $\lambda_1 \geqq \lambda_2 \geqq \cdots \geqq \lambda_n$, we may set $e^{i\theta_j} = (1 + i\lambda_j)/(1 - i\lambda_j)$, $-\pi < \theta_n \leqq \cdots \leqq \theta_1 \leqq \pi$. But

$$\prod_{j<k} \left(e^{i\theta_j} - e^{i\theta_k}\right) = \prod_{j<k} 2i\,(\lambda_j - \lambda_k) \prod_{m=1}^{n} (1 - i\lambda_m)^{-(n-1)},$$

so that

$$\prod_{j<k} |e^{i\theta_j} - e^{i\theta_k}|^2 = 2^{n(n-1)} \prod_{j<k} (\lambda_j - \lambda_k)^2 \prod_{m=1}^{n} \left(1 + \lambda_m^2\right)^{-(n-1)}.$$

Now from the equality

$$d\theta_j = \frac{2\, d\lambda_j}{\left(1 + \lambda_j^2\right)}$$

and from (3.1.14) we get

$$\dot{U} = \prod_{j<k} |e^{i\theta_j} - e^{i\theta_k}|^2\, d\theta_1 \ldots d\theta_n \cdot [\dot{U}], \tag{3.3.4}$$

where $e^{i\theta_1}, e^{i\theta_2}, \cdots, e^{i\theta_n}$ are the eigenvalues of U, and $-\pi < \theta_n \leqq \cdots \leqq \theta_1 \leqq \pi$.

3.4. **Polar coordinates for arbitrary square matrices.** We proceed to define polar coordinates for an arbitrary square matrix Z of order n. We know that any matrix Z can be represented in the form

$$Z = U \Lambda V, \tag{3.4.1}$$

where U and V are unitary matrices, and $\Lambda = [\lambda_1, \cdots, \lambda_n]$ is a diagonal matrix such that

$$\lambda_1 \geqslant \lambda_2 \geqslant \ldots \geqslant \lambda_n \geqslant 0.$$

The triple of matrices (U, Λ, V) may be called the polar coordinates of the matrix Z. The correspondence between Z and (U, Λ, V) is again not one-to-one. However, the correspondence between Z and $[\mathfrak{U}_n] \times \Lambda \times \mathfrak{U}_n$ will be one-to-one with the exception of a manifold of lower dimension. In fact, setting

$$Z = U\Lambda V = U_1\Lambda_1 V_1,$$

we obtain, first, that $\Lambda_1 = \Lambda$, since $\lambda_1^2, \lambda_2^2, \cdots, \lambda_n^2$ are the eigenvalues of the matrix $Z\bar{Z}'$. Furthermore, setting

$$U_2 = U_1^{-1} U,$$

we find

$$U_2 \Lambda^2 \bar{U}_2' = \Lambda^2.$$

Consequently, if for the matrix Z all λ_ν are different, then U_2 is a diagonal unitary matrix, and the assertion is proved for such matrices Z. On the other hand the matrices Z for which some of the λ_ν are equal (and consequently $Z\bar{Z}'$ has multiple eigenvalues) form a manifold of lower dimension.

Remark. The (real) dimension of $\mathfrak{U}_n$ is n^2, the dimension of $[\mathfrak{U}_n]$ is $n(n-1)$, the dimension of Λ is n. So the dimension of $[\mathfrak{U}_n] \times \Lambda \times \mathfrak{U}_n$ is $2n^2$, which coincides with the dimension of Z.

Now we compute the Jacobian of the transformation (3.4.1). For greater clarity we do this in several steps.

(1) It is known that any positive definite hermitian matrix H can be written in the form $T\bar{T}'$ in a unique way, where T is a triangular matrix with positive diagonal elements (zeros above the diagonal).

We set $H = (h_{jk})$

$$h_{jj} = h_j; \quad h_{jk} = h_{jk}' + ih_{jk}'', \quad j > k; \quad h_{kj} = \bar{h}_{jk},$$

where h_j, h_{jk}', h_{jk}'' are n^2 real numbers. Then

$$\dot{H} = 2^{\frac{n(n-1)}{2}} \prod_{j=1}^{n} dh_j \prod_{j>k} dh_{jk}' \, dh_{jk}''$$

is the volume element in the space of hermitian matrices.

We rewrite T in the form $T = (t_{jk})$,

$$t_{jj} = t_j; \quad t_{jk} = t_{jk}' + it_{jk}'', \quad j > k; \quad t_{jk} = 0, \qquad j < k,$$

where t_j, t_{jk}', t_{jk}'' are real numbers. Let

$$\dot{T} = \prod_{j=1}^{n} dt_j \prod_{j>k} dt_{jk}' \, dt_{jk}''$$

be the volume element in the space of all triangular matrices.

Theorem 3.4.1. *We have the equality*

$$\dot{H} = 2^{\frac{n(n+1)}{2}} t_1^{2(n-1)+1} \cdots t_{n-1}^3 t_n \cdot \dot{T}. \tag{3.4.2}$$

PROOF. We proceed by induction. For $n=1$ the theorem is clearly true since $h_1=t_1^2$. We set

$$H=\begin{pmatrix} H_1 & \bar{v}' \\ v & h_n \end{pmatrix}, \quad T=\begin{pmatrix} T_1 & 0 \\ \tau & t_n \end{pmatrix}.$$

From $H=T\bar{T}'$ we have

$$H_1=T_1\bar{T}_1', \quad v=\tau\bar{T}_1', \quad h_n=\tau\bar{\tau}'+t_n^2.$$

It is easy to see that the Jacobian $\partial(v)/\partial(\tau)=|\det T_1|^2$ and $\partial h_n/\partial t_n=2t_n$. Therefore

$$\prod_{k=1}^{n-1} dh_{nk}'\, dh_{nk}''=|\det T_1|^2\cdot\prod_{k=1}^{n-1} dt_{nk}'\, dt_{nk}''=(t_1\ldots t_{n-1})^2\prod_{k=1}^{n-1} dt_{nk}'\, dt_{nk}''.$$

By the induction hypothesis we have

$$\begin{aligned}\dot{H}&=2^{n-1}\dot{H}_1\left(\prod_{k=1}^{n-1} dh_{nk}'\, dh_{nk}''\right)dh_n\\ &=2^n(t_1\ldots t_{n-1})^2 t_n\dot{H}_1\left(\prod_{k=1}^{n-1} dt_{nk}'\, dt_{nk}''\right)dt_n\\ &=2^n(t_1\ldots t_{n-1})^2 t_n\left(2^{\frac{n(n-1)}{2}}t_1^{2(n-2)+1}\ldots t_{n-1}\right)\dot{T}\\ &=2^{\frac{n(n+1)}{2}}t_1^{2(n-1)+1}\ldots t_{n-1}^3 t_n\cdot\dot{T}.\end{aligned}$$

(2) For any nonsingular matrix Z we know from (1) that there exists a triangular matrix T of the type described in (1) such that $Z\bar{Z}'=T\bar{T}'$. This gives us the following theorem.

THEOREM 3.4.2. *Every nonsingular matrix Z can be represented in a unique fashion in the form $Z=TU$, where U is a unitary matrix and T is a triangular matrix with positive diagonal elements and zeros above the diagonal.*

PROOF. Since from $Z\bar{Z}'=T\bar{T}'$ it follows that $(Z^{-1}T)^{-1}=(\overline{Z^{-1}T})'$, we have $Z^{-1}T=U^{-1}$, i.e., $Z=TU$. Since a unitary triangular matrix with positive diagonal elements must be equal to I, we obtain, using the relations

$$Z=TU=T_1U_1$$

that $T=T_1$ and $U=U_1$.

THEOREM 3.4.3. *The volume element of Z in the parametric form of Theorem 3.4.2 is given by*

$$\dot{Z}=2^{-\frac{n(n-1)}{2}}t_1^{2(n-1)+1}\dots t_{n-1}^3t_n\dot{T}\dot{U}, \tag{3.4.3}$$

where $\dot{U}$ is the volume element of the space of unitary matrices.

PROOF. From the equality $Z=TU$ we obtain

$$dZ=dT\cdot U+T\cdot dU.$$

Interchanging rows and columns and taking adjoints we also have

$$d\bar{Z}'=U^{-1}d\bar{T}'+dU^{-1}\cdot\bar{T}'=U^{-1}d\bar{T}'-U^{-1}dU\cdot U^{-1}\bar{T}'.$$

We set $\delta U=dU\cdot U^{-1}$ and $dP=dZ\cdot U^{-1}$, $dQ=Ud\bar{Z}'$. Then

$$\begin{aligned} dP&=dT+T\delta U,\\ dQ&=d\bar{T}'-\delta U\cdot\bar{T}' \end{aligned} \tag{3.4.4}$$

(δU here is not the same as in §3.1, but has the same properties). Writing $\delta U=(dv_{jk})$ and equating the corresponding elements in (3.4.4) we have

$$dp_{jk}=dt_{jk}+\sum_{s=1}^{j}t_{js}\,dv_{sk},\qquad j>k, \tag{3.4.5}$$

$$dq_{jk}=d\bar{t}_{kj}-\sum_{s=1}\bar{t}_{ks}\,dv_{js},\qquad j<k, \tag{3.4.6}$$

$$dp_{jj}=dt_j+\sum_{s=1}^{j}t_{js}\,dv_{sj}, \tag{3.4.7}$$

$$dp_{jk}=\sum_{s=1}^{j}t_{js}\,dv_{sk},\qquad j<k, \tag{3.4.8}$$

$$dq_{jk}=-\sum_{s=1}^{k}\bar{t}_{ks}\,dv_{js},\qquad j>k, \tag{3.4.9}$$

$$dq_{jj}=dt_j-\sum_{s=1}^{j}\bar{t}_{js}\,dv_{js}. \tag{3.4.10}$$

Subtracting (3.4.10) from (3.4.7) we get

$$d(p_{jj}-q_{jj})=2t_j\,dv_{jj}+\sum_{s=1}^{j-1}(t_{js}\,dv_{sj}+\bar{t}_{js}\,dv_{js}). \tag{3.4.11}$$

Now we compute the following Jacobian J:

$$\left|\frac{\partial\{p_{jk}\,(j>k), q_{jk}\,(j<k), p_{jj}\,(1\leqslant j\leqslant n), p_{jk}\,(j<k), q_{jk}\,(j>k), p_{jj}-q_{jj}\,(1\leqslant j\leqslant n)\}}{\partial\{t_{jk}\,(j>k),\ \bar{t}_{kj}\,(j<k),\ t_j,\ v_{jk}\,(1\leqslant j,k\leqslant n)\}}\right|$$

from (3.4.8), (3.4.9), and (3.4.11) we have, owing to the absence of dt_{jk}, $d\bar{t}_{jk}$, dt_j,

$$J=\left|\frac{\partial\{p_{jk}\,(j<k),\ q_{jk}\,(j>k),\ p_{jj}-q_{jj}\,(1\leqslant j\leqslant n)\}}{\partial\{v_{jk}\,(1\leqslant j,k\leqslant n)\}}\right|.$$

We note that in (3.4.8) only the dv_{pq} $(p<q)$ occur and in (3.4.9) only the dv_{pq} $(p>q)$. Therefore dp_{jk} $(j<k)$ and dq_{jk} $(j>k)$ are linear combinations of dv_{pq} $(p<q)$ and dv_{pq} $(p>q)$, respectively. The matrices of these transformations are triangular, and their determinant is equal to $t_1^{n-1}t_2^{n-2}\cdots t_{n-1}$. Therefore

$$J=2^n\cdot t_1^{2(n-1)+1}\cdots t_{n-1}^3 t_n.$$

Since

$$\left|\frac{\partial(z,\bar{z})}{\partial(x,y)}\right|=2,$$

we have

$$\left|\frac{\partial(p_{jk},q_{jk})}{\partial\left(t'_{jk},t''_{jk},t_j,v_{jk}\right)}\right|=2^{n(n-1)}\left|\frac{\partial(p_{jk},q_{jk})}{\partial(t_{jk},\bar{t}_{jk},t_j,v_{jk})}\right|.$$

It follows immediately that

$$\left|\frac{\partial(P,Q)}{\partial(T,V)}\right|=2^{n^2}t_1^{2(n-1)+1}\ \cdots\ t_{n-1}^3\cdot t_n.$$

Furthermore, since

$$\left|\frac{\partial(Z,\bar{Z})}{\partial(P,Q)}\right|=1,\quad\left|\frac{\partial(Z,\bar{Z})}{\partial(X,Y)}\right|=2^{n^2},$$

we finally obtain

$$\dot{Z}=2^{-\frac{n(n-1)}{2}}t_1^{2(n-1)+1}\ \cdots\ t_{n-1}^3 t_n\cdot\dot{T}\dot{U}.$$

(3) The volume element in polar coordinates.

THEOREM 3.4.4. *The relation between the volume elements in the two coordinate systems Z and $[\mathfrak{U}_n]\times\Lambda\times\mathfrak{U}_n$ is given by the formula*

$$\dot{Z} = 2^{-n^2} D^2(\lambda_1, \ldots, \lambda_n)\, d\lambda_1 \ldots d\lambda_n \dot{U}\,[\dot{U}]. \tag{3.4.12}$$

PROOF. Setting $Z\bar{Z}' = T\bar{T}' = H$, from Theorem 3.4.3 we have

$$\dot{Z} = 2^{-\frac{n(n-1)}{2}} t_1^{2(n-1)+1} \ldots t_{n-1}^3 t_n \dot{T} \dot{U},$$

and, consequently, by Theorem 3.4.1 we have

$$\dot{Z} = 2^{-n^2} \dot{H} \dot{U}.$$

Hence, by (3.3.2)

$$\dot{Z} = 2^{-n^2} D^2(\lambda_1, \ldots, \lambda_n)\, d\lambda_1 \ldots d\lambda_n \cdot \dot{U}\,[\dot{U}].$$

3.5. **Polar coordinates for symmetric matrices.** If Z is a symmetric matrix with complex elements, then as it has been shown by the author (L. K. Hua [1]), it can be represented in the form

$$Z = U\Lambda U', \tag{3.5.1}$$

where U is unitary and $\Lambda = [\lambda_1, \cdots, \lambda_n]$ is a diagonal matrix with $\lambda_1 \geqq \lambda_2 \geqq \cdots \geqq \lambda_n \geqq 0$. We denote by $\{\mathfrak{U}_n\}$ the set of left cosets of the group $\mathfrak{U}_n$ of unitary matrices with respect to the subgroup of diagonal matrices of the form $[\pm 1, \pm 1, \cdots, \pm 1]$. It is easy to show that the correspondence between Z and $\{\mathfrak{U}_n\} \times \Lambda$ is one-to-one for almost all matrices.

Differentiating (3.5.1) we get

$$dZ = dU\Lambda U' + Ud\Lambda \cdot U' + U\Lambda dU'$$

$$\bar{U}' dZ \bar{U} = \delta U \Lambda + d\Lambda + \Lambda \delta U'.$$

(Here we have written $\delta U = \bar{U}' dU$.) Hence we have

$$\begin{aligned}\mathrm{Sp}(dZ\, d\bar{Z}') &= \mathrm{Sp}(\bar{U}'\, dZ \cdot U\bar{U}' \cdot d\bar{Z}' \cdot U)\\ &= \mathrm{Sp}\,\{(\delta U \cdot \Lambda + d\Lambda + \Lambda\, \delta U')(\Lambda\, \delta\bar{U}' + d\Lambda + \delta\bar{U} \cdot \Lambda)\}\\ &= \mathrm{Sp}(d\Lambda \cdot d\Lambda) + \mathrm{Sp}\,\{(\delta U \cdot \Lambda + \Lambda\, \delta U')(\Lambda\, \delta \bar{U}' + \delta\bar{U} \cdot \Lambda)\}.\end{aligned}$$

We set

$$\delta U \cdot \Lambda + \Lambda \delta U' = (dg_{jk}), \qquad (dg_{jk} = dg_{kj}).$$

Then

$$\mathrm{Sp}(dZ \cdot d\bar{Z}') = \sum_{j=1}^n d\lambda_j^2 + \sum_{j=1}^n |dg_{jj}|^2 + 2\sum_{j<k} |dg_{jk}|^2,$$

where

$$dg_{jk} = \lambda_k \delta u_{jk} + \lambda_j \delta u_{kj}, \qquad j<k,$$
$$dg_{jj} = 2i\lambda_j \delta u_{jj}.$$

In order to determine the volume element $\{\dot{U}\}$ of the manifold $\{\mathfrak{U}_n\}$, we set $\delta u_{jk} = \delta u'_{jk} + i\delta u''_{jk}$. We have

$$\{\dot{U}\} = 2^{\frac{n(n-1)}{2}} \prod_{j=1}^{n} \delta u''_{jj} \prod_{j<k} \delta u'_{jk} \cdot \delta u''_{jk}.$$

Consequently,

$$\dot{Z} = 2^n \prod_{j<k} |\lambda_j^2 - \lambda_k^2| \lambda_1 \ldots \lambda_n \, d\lambda_1 \ldots d\lambda_n \{\dot{U}\}. \tag{3.5.2}$$

Now we consider polar coordinates for real-symmetric matrices. Any real-symmetric matrix T can be represented in the form

$$T = \Gamma \Lambda \Gamma', \tag{3.5.3}$$

where Γ is a real orthogonal matrix of determinant $+1$ and

$$\Lambda = [\lambda_1, \cdots, \lambda_n], \qquad \lambda_1 \geqq \lambda_2 \geqq \cdots \geqq \lambda_n.$$

It is easy to see that there is in general a one-to-one correspondence between real symmetric matrices and $\{O\} \times \Lambda$, where $\{O\}$ is the set of left cosets of the group of orthogonal matrices with determinant $+1$ with respect to the subgroup of all diagonal matrices of the form $[\pm 1, \cdots, \pm 1]$.

Differentiating (3.5.3) we obtain

$$\Gamma' dT \Gamma = \delta\Gamma \cdot \Lambda + d\Lambda - \Lambda \delta\Gamma.$$

Here $\delta\Gamma = \Gamma^{-1} d\Gamma$ is a skew-symmetric matrix such that

$$\mathrm{Sp}(dT \cdot dT') = \mathrm{Sp}\{(\delta\Gamma \cdot \Lambda - \Lambda\delta\Gamma)(\delta\Gamma \cdot \Lambda - \Lambda\delta\Gamma)\} + \mathrm{Sp}(d\Lambda \cdot d\Lambda').$$

We set $[\dot{O}] = \prod_{1 \leq i<j \leq n} \delta\gamma_{ij}$. By the same considerations as above we find that

$$\dot{T} = 2^{\frac{n(n-1)}{2}} \prod_{i<j} |\lambda_i - \lambda_j| \, d\lambda_1 \ldots d\lambda_n [\dot{O}]. \tag{3.5.4}$$

We denote by $\mathfrak{S}$ the set of all symmetric unitary matrices S. We shall find the volume element of $\mathfrak{S}$ invariant under transformations of the form

$$S_1 = USU', \tag{3.5.5}$$

where U is a unitary matrix.

We set

$$S=(I+iT)(I-iT)^{-1} \tag{3.5.6}$$

and we call the matrix T the parameter of the matrix S. Since S is unitary, it follows that T is hermitian; since S is symmetric, it follows that T is also symmetric. So $T=T'=\overline{T'}$, i.e., T is a real-symmetric matrix. The formula inverse to (3.5.6) has the form

$$T=i(I-S)(I+S)^{-1}. \tag{3.5.7}$$

The formulas (3.5.6) and (3.5.7) establish a correspondence between S and T which is one-to-one for almost all matrices. (The exceptional matrices form a manifold of dimension less than $n(n+1)/2$).

If the matrix S is transformed by (3.5.5) into the matrix S_1 then the matrix T is transformed into some matrix T_1. We find the relation between T and T_1. We split the matrix U into its real and imaginary parts

$$U=A+Bi. \tag{3.5.8}$$

Since $U\overline{U'}=I$, the real matrices A and B satisfy the relations

$$AA'+BB'=I, \qquad AB'=BA'. \tag{3.5.9}$$

Since

$$\begin{aligned} S_1 &= USU' = (A+Bi)(I+iT)(I-iT)^{-1}\overline{U}^{-1} \\ &= [(A-BT)+(B+AT)i](I-iT)^{-1}(A-iB)^{-1} \\ &= [I+i(B+AT)(A-BT)^{-1}][I-i(B+AT)(A-BT)^{-1}]^{-1}, \end{aligned}$$

we have

$$T_1=(AT+B)(-BT+A)^{-1}. \tag{3.5.10}$$

From (3.5.9) it follows in an obvious way that

$$(AT+B)(-BT+A)^{-1}=(-TB'+A')^{-1}(TA'+B'). \tag{3.5.11}$$

Differentiating (3.5.10) and using the relation (3.5.11) we obtain

$$\begin{aligned} dT_1 &= A\,dT\cdot(-BT+A)^{-1}+(AT+B)(-BT+A)^{-1}B\,dT \\ &\quad\times(-BT+A)^{-1}=A\,dT\cdot(-BT+A)^{-1}+(-TB'+A')^{-1} \\ &\quad\times(TA'+B')B\,dT\cdot(-BT+A)^{-1}=(-TB'+A')^{-1} \\ &\quad\times[(-TB'+A')A+(TA'+B')B]\,dT\cdot(-BT+A)^{-1} \\ &= [(-BT+A)']^{-1}\,dT\cdot(-BT+A)^{-1}. \end{aligned} \tag{3.5.12}$$

With the help of (3.5.9) and (3.5.11) we find

$$\begin{aligned} I+T_1^2 &= I+(-TB'+A')^{-1}(TA'+B')(AT+B)(-BT+A)^{-1} \\ &= [(-BT+A)']^{-1}[(-TB'+A')(-BT+A) \\ &\quad +(TA'+B')(AT+B)](-BT+A)^{-1} \\ &= [(-BT+A)']^{-1}(I+T^2)(-BT+A)^{-1}. \end{aligned} \tag{3.5.13}$$

From (3.5.12) we obtain

$$\dot{T}_1 = \{\det(-BT+A)\}^{-(n+1)}\dot{T},$$

and from (3.5.12) we have

$$\det(I+T_1^2) = \det(I+T^2)\,\{\det(-BT+A)\}^{-2}.$$

Therefore

$$\{\det(I+T_1^2)\}^{-\frac{n+1}{2}}\cdot\dot{T}_1 = \{\det(I+T^2)\}^{-\frac{n+1}{2}}\cdot\dot{T}. \tag{3.5.14}$$

By a method analogous to that of §3.1 we find

$$\dot{S} = 2^{\frac{n(n+1)}{2}}\cdot\{\det(I+T^2)\}^{-\frac{n+1}{2}}\cdot\dot{T}, \tag{3.5.15}$$

where

$$\dot{T} = 2^{n(n-1)/4}\prod_{j\le k} dt_{jk.}$$

It is clear that $\dot{S}=\dot{S}_1$, i.e., this is the invariant volume element. On the basis of Theorem 2.1.1 we find that the total volume of $\mathfrak{S}$ is equal to

$$\begin{aligned} \int_S \dot{S} &= 2^{\frac{n(n+1)}{2}}\int_T \{\det(I+T^2)\}^{-\frac{n+1}{2}}\cdot\dot{T} \\ &= 2^n\cdot 2^{\frac{3n(n-1)}{4}}\cdot\pi^{\frac{n(n+1)}{4}}\frac{\Gamma\left(\frac{1}{2}\right)}{\Gamma\left(\frac{n+1}{2}\right)}\cdot\prod_{\nu=1}^{n-1}\frac{\Gamma\left(\frac{n}{2}-\frac{\nu}{2}+1\right)}{\Gamma(n-\nu+1)}. \end{aligned}$$

Using (3.5.4) we transform the expression for $\dot{S}$ into the form

$$\dot{S} = 2^n\cdot 2^{\frac{3n(n-1)}{4}}\cdot\prod_{1\leqslant i<j\leqslant n}|\lambda_i-\lambda_j|\prod_{\nu=1}^{n}(1+\lambda_\nu^2)^{-\frac{n+1}{2}}\cdot d\lambda_1\ldots d\lambda_n\cdot\{\dot{O}\}. \tag{3.5.16}$$

We set $e^{i\theta_\nu}=(1+i\lambda_\nu)(1-i\lambda_\nu)^{-1}$. Since $\lambda_1\geqq\lambda_2\geqq\cdots\geqq\lambda_n$, we may assume that $\pi\geqq\theta_1\geqq\theta_2\geqq\cdots\geqq\theta_n\geqq-\pi$. It is easy to verify that

$$e^{i\theta_\nu}-e^{i\theta_\mu} = \frac{2i(\lambda_\nu-\lambda_\mu)}{(1-i\lambda_\nu)(1-i\lambda_\mu)},$$

and therefore

$$\left| e^{i\theta_\nu} - e^{i\theta_\mu} \right| = \frac{2\,|\lambda_\nu - \lambda_\mu|}{(1+\lambda_\nu^2)^{\frac{1}{2}}(1+\lambda_\mu^2)^{\frac{1}{2}}},$$

whence we obtain

$$\prod_{1\leqslant\nu<\mu\leqslant n} \left| e^{i\theta_\nu} - e^{i\theta_\mu} \right| = \frac{2^{\frac{n(n-1)}{2}} \prod_{\nu<\mu} |\lambda_\nu - \lambda_\mu|}{\prod_{\nu=1}^{n} (1+\lambda_\nu^2)^{\frac{n-1}{2}}}. \qquad (3.5.17)$$

On the other hand we have

$$d\theta_\nu = \left(\frac{1}{1+i\lambda_\nu} + \frac{1}{1-i\lambda_\nu} \right) d\lambda_\nu = \frac{2}{1+\lambda_\nu^2}\, d\lambda_\nu. \qquad (3.5.18)$$

Combining (3.5.16), (3.5.17) and (3.5.18) we get still another expression for $\dot{S}$,

$$\dot{S} = 2^{\frac{n(n-1)}{2}} \prod_{1\leqslant\nu<\mu\leqslant n} \left| e^{i\theta_\nu} - e^{i\theta_\mu} \right| d\theta_1 \ldots d\theta_n \{\dot{O}\}. \qquad (3.5.19)$$

REMARK. From the considerations in this section the following result follows: Any symmetric unitary matrix can be represented[11] in the form

$$S = \Gamma\Lambda\Gamma',$$

where Γ is a real orthogonal matrix with determinant 1, and $\Lambda = [e^{i\theta_1}, e^{i\theta_2}, \cdots, e^{i\theta_n}]$ with $\pi \geqq \theta_1 \geqq \theta_2 \geqq \cdots \geqq \theta_n \geqq -\pi$.

3.6. **Polar coordinates for skew-symmetric matrices.** The author has proved (L. K. Hua [1]) that any skew-symmetric matrix Z with complex entries can be represented in the form

$$Z = UMU', \qquad (3.6.1)$$

where U is a unitary matrix and

$$M = \begin{pmatrix} 0 & \lambda_1 \\ -\lambda_1 & 0 \end{pmatrix} \dot{+} \begin{pmatrix} 0 & \lambda_2 \\ -\lambda_2 & 0 \end{pmatrix} \dot{+} \cdots, \quad \lambda_1 \geqslant \lambda_2 \geqslant \cdots \geqslant \lambda_\nu \geqslant 0.$$

---

[11] Note by the editor of the Russian translation. For the proof it suffices to use (3.5:6) and (3.5.3).

Here the last term is equal to $\begin{pmatrix} 0 & \lambda_\nu \\ -\lambda_\nu & 0 \end{pmatrix}$, in case n is even; in the case of odd n the last terms are $\begin{pmatrix} 0 & \lambda_\nu \\ -\lambda_\nu & 0 \end{pmatrix} \dotplus 0 \quad \left(\nu = \left[\frac{n}{2}\right]\right)$.

We should also mention that for odd n any skew-symmetric unitary matrix can be represented in the form

$$K = \Gamma F \Gamma', \tag{3.6.2}$$

where Γ is a real orthogonal matrix with determinant 1, and

$$F = \begin{pmatrix} 0 & e^{i\theta_1} \\ -e^{i\theta_1} & 0 \end{pmatrix} \dotplus \begin{pmatrix} 0 & e^{i\theta_2} \\ -e^{i\theta_2} & 0 \end{pmatrix} \dotplus \cdots, \quad \pi \geqslant \theta_1 \geqslant \theta_2 \geqslant \cdots \geqslant \theta_\nu \geqslant 0.$$

For even n any skew-symmetric matrix can again be represented in the form (3.6.2), but only under the condition that the determinant of Γ may be $+1$ or -1. We may mention also that using real instead of complex parameters in (3.6.2) we get a representation of the manifold of skew-symmetric matrices which has real dimension $n(n-1)$.

We denote by $\mathfrak{K}$ the set of matrices representable in the form (3.6.2). If $\Gamma F \Gamma' = \Gamma_1 F_1 \Gamma_1'$, then $F = F_1$, and if the numbers $e^{i\theta_1}, \cdots, e^{i\theta_n}$ are all different, then $\Gamma = \Gamma_1 \Delta$, where

$$\Delta = \begin{pmatrix} \cos\delta_1 & \sin\delta_1 \\ -\sin\delta_1 & \cos\delta_1 \end{pmatrix} \dotplus \begin{pmatrix} \cos\delta_2 & \sin\delta_2 \\ -\sin\delta_2 & \cos\delta_2 \end{pmatrix} \dotplus \cdots. \tag{3.6.3}$$

The last terms of this sum are of one of the two forms

$$\begin{pmatrix} \cos\delta_\nu & \sin\delta_\nu \\ -\sin\delta_\nu & \cos\delta_\nu \end{pmatrix} \quad \text{or} \quad \begin{pmatrix} \cos\delta_\nu & \sin\delta_\nu \\ -\sin\delta_\nu & \cos\delta_\nu \end{pmatrix} \dotplus 1$$

($\nu = [n/2]$), depending on whether n is even or odd.

We denote by Γ the group of orthogonal matrices with determinant $+1$, by Δ the group of matrices of the form (3.6.3), and by $\Sigma = \Gamma/\Delta$ the set of left cosets of Γ with respect to the subgroup Δ. Then the correspondence between $\mathfrak{K}$ and $\Sigma \times F$ will be one-to-one for almost all matrices.

Differentiating (3.6.2) we obtain

$$dK = d\Gamma \cdot F\Gamma' + \Gamma dF \cdot \Gamma' + \Gamma F d\Gamma'.$$

We set $\delta\Gamma = \Gamma' d\Gamma$. Then $\Gamma' dK \Gamma = \delta\Gamma F + dF - F\delta\Gamma$ and

$$\Gamma' d\overline{K}' \Gamma = -\overline{F}' \delta\Gamma + \delta\Gamma \overline{F}' + d\overline{F}'.$$

Consequently,

$$\begin{aligned}\operatorname{Sp}(dK\, d\bar{K}') &= \operatorname{Sp}(\Gamma'\, dK\,\Gamma \cdot \Gamma'\, d\bar{K}' \cdot \Gamma)\\ &= \operatorname{Sp}\{(\delta\Gamma\, F - F\,\delta\Gamma)(\delta\Gamma\,\bar{F}' - \bar{F}'\,\delta\Gamma)\} + \operatorname{Sp}\{(\delta\Gamma\, F - F\,\delta\Gamma)\, d\bar{F}'\} \\ &\quad + \operatorname{Sp}\{dF\,(-\bar{F}'\,\delta\Gamma + \delta\Gamma\bar{F}')\} + \operatorname{Sp}(dF\, d\bar{F}').\end{aligned} \tag{3.6.4}$$

Since

$$dF = iF[d\theta_1, d\theta_1, d\theta_2, d\theta_2, \cdots],$$

we have

$$\operatorname{Sp}(dF\, d\bar{F}') = 2(d\theta_1^2 + d\theta_2^2 + \cdots + d\theta_\nu^2) \qquad (\nu = [n/2]). \tag{3.6.5}$$

Besides, since $dF\bar{F}' = \bar{F}' dF$, we have

$$\operatorname{Sp}\{(\delta\Gamma F - F\delta\Gamma) d\bar{F}'\} = \operatorname{Sp}(\delta\Gamma\cdot F\cdot d\bar{F}') - \operatorname{Sp}(\delta\Gamma\cdot d\bar{F}'\cdot F) = 0 \tag{3.6.6}$$

and

$$\operatorname{Sp}\{dF(-\bar{F}'\delta\Gamma + \delta\Gamma\bar{F}')\} = 0. \tag{3.6.7}$$

Furthermore,

$$\operatorname{Sp}\{(\delta\Gamma F - F\delta\Gamma)(\delta\Gamma\bar{F}' - \bar{F}'\delta\Gamma)\} = 2\operatorname{Sp}(\delta\Gamma F\delta\Gamma\bar{F}') - 2\operatorname{Sp}((\delta\Gamma)^2\bar{F}'F). \tag{3.6.8}$$

Consequently, from (3.6.4) and (3.6.8) we get

$$\operatorname{Sp}(dK\cdot d\bar{K}') = 2\operatorname{Sp}(\delta\Gamma\cdot F\cdot\delta\Gamma\cdot\bar{F}') - 2\operatorname{Sp}((\delta\Gamma)^2\bar{F}'F) + 2\sum_{\alpha=1}^{\nu} d\theta_\alpha^2. \tag{3.6.9}$$

Formally we have a quadratic form of the $n(n-1)/2+\nu$ differentials $\delta\gamma_{\alpha\beta}$, $1 \leqq \alpha < \beta \leqq n$ and $d\theta_1, d\theta_2, \cdots, d\theta_\nu$, but in fact $\delta\gamma_{12}, \delta\gamma_{34}, \cdots$ do not occur. In other words, we have a quadratic form of $n(n-1)/2+\nu-\nu = n(n-1)/2$ differentials only. Now we compute its discriminant (i.e., the volume element).

THEOREM 3.6.1. *The volume element in polar coordinates of the manifold* $\mathfrak{K}$ *is equal to*

$$\dot{K} = \varkappa_n \cdot \prod_{1\leqslant\alpha<\beta\leqslant\nu} \sin^2(\theta_\beta - \theta_\alpha) \cdot \prod_{\alpha=1}^{\nu} d\theta_\alpha \cdot \dot{\Sigma}, \tag{3.6.10}$$

where

$$\varkappa_n = \begin{cases} 2^{2\nu(\nu-1)+\frac{\nu}{2}}, & n = 2\nu \\ 2^{2\nu(\nu-1)+\frac{3\nu}{2}}, & n = 2\nu+1, \end{cases}$$

and $\dot{\Sigma}$ is the volume element in the coset space $\Sigma = \Gamma/\Delta$. More precisely, $\dot{\Sigma}$ is equal to the product of all $\delta\gamma_{ij}$, $1 \leqq i < j \leqq n$, with the exception of $\delta\gamma_{12}, \delta\gamma_{34}, \cdots$ (the number of which equals $\nu = [n/2]$).

PROOF. Let first $n = 2\nu$ be an even number. We decompose F and $\delta\Gamma$ into blocks of two-by-two matrices

$$\delta\Gamma = (\delta\Gamma_{\alpha\beta})_1^\nu, \qquad (\delta\Gamma_{\alpha\beta})' = -(\delta\Gamma_{\beta\alpha})$$

and

$$F = F_1 \dotplus F_2 \dotplus \dots \dotplus F_\nu,$$

where

$$F_\alpha = e^{i\theta_\alpha} \cdot F_0, \qquad \overline{F}'_\alpha = -e^{-i\theta_\alpha} F_0, \qquad F_0 = \begin{pmatrix} 0 & 1 \\ -1 & 0 \end{pmatrix}.$$

Substituting into (3.6.9) we get

$$\begin{aligned} &\operatorname{Sp}(dK \cdot d\overline{K}') \\ &= 2 \sum_{\alpha,\beta=1}^{\nu} \operatorname{Sp}\left(\delta\Gamma_{\alpha\beta} \cdot F_\beta \cdot \delta\Gamma_{\beta\alpha} \cdot \overline{F}'_\alpha\right) - 2 \sum_{\alpha,\beta=1}^{\nu} \operatorname{Sp}(\delta\Gamma_{\alpha\beta}\,\delta\Gamma_{\beta\alpha}) + 2 \sum_{\alpha=1}^{\nu} d\theta_\alpha^2 \\ &= -2 \sum_{\alpha=1}^{\nu} \{\operatorname{Sp}(\delta\Gamma_{\alpha\alpha} \cdot F_0 \cdot \delta\Gamma_{\alpha\alpha} \cdot F_0) + \operatorname{Sp}((\delta\Gamma_{\alpha\alpha})^2)\} \\ &+ 4 \sum_{1 \leqslant \alpha < \beta \leqslant \nu} \{\operatorname{Sp}\left(\delta\Gamma_{\alpha\beta} \cdot \delta\Gamma'_{\alpha\beta}\right) + \cos(\theta_\beta - \theta_\alpha) \operatorname{Sp}\left(\delta\Gamma_{\alpha\beta} \cdot F_0 \cdot \delta\Gamma'_{\alpha\beta} \cdot F_0\right)\} \\ &+ 2 \sum_{\alpha=1}^{\nu} d\theta_\alpha^2. \end{aligned} \tag{3.6.11}$$

First we set

$$\delta\Gamma_{\alpha\alpha} = \begin{pmatrix} 0 & \rho \\ -\rho & 0 \end{pmatrix},$$

whence we have

$$\operatorname{Sp}(\delta\Gamma_{\alpha\alpha} \cdot F_0 \cdot \delta\Gamma_{\alpha\alpha} \cdot F_0) + \operatorname{Sp}((\delta\Gamma_{\alpha\alpha})^2) = 0. \tag{3.6.12}$$

Furthermore, we set

$$\delta\Gamma_{\alpha\beta} = \begin{pmatrix} a & b \\ c & d \end{pmatrix}.$$

Then

$$\begin{aligned} &4 \operatorname{Sp}\left(\delta\Gamma_{\alpha\beta} \cdot \delta\Gamma'_{\alpha\beta}\right) + 4 \cos(\theta_\beta - \theta_\alpha) \operatorname{Sp}\left(\delta\Gamma_{\alpha\beta} \cdot F_0 \cdot \delta\Gamma'_{\alpha\beta} \cdot F_0\right) \\ &\qquad = 4\{a^2 + b^2 + c^2 + d^2 + 2\cos(\theta_\beta - \theta_\alpha)(bc - ad)\}. \end{aligned} \tag{3.6.13}$$

We have obtained a quadratic form of the four variables a, b, c, and d. Its discriminat equals

$$4^4 \sin^4(\theta_\beta - \theta_\alpha).$$

From (3.6.12) and (3.6.13) we see that (3.6.11) is a quadratic form of $4\frac{1}{2}\nu(\nu-1)+\nu = n(n-1)/2$ variables, and its discriminat is equal to

$$2^\nu 4^{2\nu(\nu-1)} \prod_{1\leqslant\alpha<\beta\leqslant\nu} \sin^4(\theta_\beta - \theta_\alpha).$$

From this it follows that the volume element of $\mathfrak{K}$ is equal to

$$2^{\frac{\nu}{2}+2\nu(\nu-1)} \prod_{1\leqslant\alpha<\beta\leqslant\nu} \sin^4(\theta_\beta - \theta_\alpha) \prod_{\alpha=1}^{\nu} d\theta_\alpha\, \dot{\Sigma}. \tag{3.6.14}$$

Now we consider the case where $n = 2\nu+1$ is an odd number. We represent $\delta\Gamma$ and F in the form

$$\delta\Gamma = \begin{pmatrix} \delta\Gamma_1 & \delta v \\ -\delta v' & 0 \end{pmatrix}, \quad F = F_1 \dot{+} 0.$$

Substituting into (3.6.9) we obtain

$$\operatorname{Sp}(dK\, d\overline{K}') = 2\operatorname{Sp}\left\{\begin{pmatrix} \delta\Gamma_1 F_1 & 0 \\ -\delta v' F_1 & 0 \end{pmatrix}\begin{pmatrix} \delta\Gamma_1 \overline{F}_1' & 0 \\ -\delta v' \overline{F}_1' & 0 \end{pmatrix}\right\}$$

$$-2\operatorname{Sp}\left\{\begin{pmatrix} \delta\Gamma_1\,\delta\Gamma_1 - \delta v\,\delta v' & \delta\Gamma_1\,\delta v \\ -\delta v'\,\delta\Gamma_1 & -\delta v'\,\delta v \end{pmatrix}\begin{pmatrix} I & 0 \\ 0 & 0 \end{pmatrix}\right\} + 2\sum_{\alpha=1}^{\nu} d\theta_\alpha^2$$

$$= 2\operatorname{Sp}(\delta\Gamma_1 F_1 \delta\Gamma_1 \overline{F}_1') - 2\operatorname{Sp}((\delta\Gamma_1)^2) + 2\operatorname{Sp}(\delta v \cdot \delta v') + 2\sum_{\alpha=1}^{\nu} d\theta_\alpha^2.$$

Using the result obtained for even n, we find that for odd n the volume element of $\mathfrak{K}$ equals

$$2^\nu\, 2^{2\nu(\nu-1)+\frac{\nu}{2}} \prod_{1\leqslant\alpha<\beta\leqslant\nu} \sin^2(\theta_\beta - \theta_\alpha) \prod_{\alpha=1}^{\nu} d\theta_\alpha \cdot \dot{\Sigma}. \tag{3.6.15}$$

Now we consider the set of skew-symmetric matrices of the form

$$K = UDU', \tag{3.6.16}$$

where U is a unitary matrix, and

$$D = \begin{pmatrix} 0 & 1 \\ -1 & 0 \end{pmatrix} \dot{+} \begin{pmatrix} 0 & 1 \\ -1 & 0 \end{pmatrix} \dot{+} \cdots$$

[the last term of the sum is $\begin{pmatrix} 0 & 1 \\ -1 & 0 \end{pmatrix}$ or 0, depending on whether n is even or odd]. For even n the set of matrices of the form (3.6.16) coincides with the set of matrices of the form (3.6.2). For odd n, however, the situation is different. It is clear that the set of matrices of the form (3.6.2) has the dimension $\frac{1}{2}n(n-1)$, and using (3.6.16) it is easy to show that the dimension of the set of matrices of this form is $\frac{1}{2}n(n+1)-1$.

3.7. **The volume of the space of real orthogonal matrices and applications.** Concluding the chapter, we find the volume element and the volume of the set of real orthogonal matrices, using the same method as in the preceding sections.

First we consider the orthogonal group O_n, i.e., the set of all real matrices T satisfying the condition

$$TT' = I. \tag{3.7.1}$$

It is clear that $\det T = \pm 1$. We denote by O_n^+ the group of matrices T of determinant $+1$. To every matrix T there corresponds a matrix K given by the formula

$$K = (I-T)(I+T)^{-1}. \tag{3.7.2}$$

The case where $\det(I+T)=0$ has to be excluded here. Note that now we cannot say "for almost all matrices the determinant of $I+T$ is different from 0". In fact, for every matrix T with determinant equal to -1, we have

$$\det(I+T) = \det(TT'+T) = \det T \cdot \det(T'+I) = -\det(I+T),$$

i.e., $\det(I+T)=0$. Therefore, we have to restrict ourselves to matrices in O_n^+. It can be shown that the set of matrices in O_n^+, for which $\det(I+T)=0$, is a manifold of lower dimension.

From (3.7.1) and (3.7.2) it follows immediately that

$$K' = -K. \tag{3.7.3}$$

Inverting (3.7.2) we obtain at once

$$T = (I-K)(I+K)^{-1}. \tag{3.7.4}$$

Whence it also follows that $\det T = 1$, since

$$\det(I-K) = \det(I-K') = \det(I+K).$$

Differentiating (3.7.4) we find

$$dT = -2(I+K)^{-1}dK(I+K)^{-1},$$

Consequently,

$$\mathrm{Sp}(dT\,dT') = -4\,\mathrm{Sp}\{dK\cdot(I-K^2)^{-1}\cdot dK\cdot(I-K^2)^{-1}\}. \qquad (3.7.5)$$

We set

$$dK = (dk_{ij}), \qquad dk_{ij} = -dk_{ji},$$
$$(I-K^2)^{-1} = (u_{st}), \qquad u_{st} = u_{ts}.$$

Then (3.7.5) gives

$$\mathrm{Sp}\,(dT\cdot dT') = -8\sum_{i<j}\sum_{s<t}(u_{js}u_{it} - u_{is}u_{jt})\,dk_{ij}\,dk_{st},$$

whence we get the expression for the volume element

$$\dot{T} = 2^{\frac{n(n-1)}{2}}\{\det(I-K^2)\}^{-\frac{n-1}{2}}\dot{K}, \qquad (3.7.6)$$

where

$$\dot{K} = 2^{\frac{n(n-1)}{4}}\prod_{i<j} dk_{ij}.$$

It follows that the volume of the orthogonal group O_n^+ is equal to

$$\int_T \dot{T} = 2^{\frac{n(n-1)}{2}}\int_K \{\det(I-K^2)\}^{-\frac{n-1}{2}}\cdot\dot{K}.$$

From Theorem 2.1.4 we obtain

$$\int_T \dot{T} = 2^{\frac{3n(n-1)}{4}}\pi^{\frac{n(n-1)}{4}}\cdot\prod_{\nu=2}^{n}\frac{\Gamma\left(\frac{\nu-1}{2}\right)}{\Gamma(\nu-1)}. \qquad (3.7.7)$$

Now we are able to find the volume of the manifold $\mathfrak{K}$.

THEOREM 3.7.1. *We have the equalities*

$$\underset{\pi>\theta_1>\cdots>\theta_\nu>0}{\int\cdots\int}\ \prod_{1\leqslant\alpha<\beta\leqslant\nu}\sin^2(\theta_\beta-\theta_\alpha)\,d\theta_1\ldots d\theta_\nu = 2^{-\nu^2}\cdot(2\pi)^\nu \qquad (3.7.8)$$

and

$$\underset{\pi>\theta_1>\cdots>\theta_\nu>0}{\int\cdots\int}\ \prod_{1\leqslant\alpha<\beta\leqslant\nu}(\cos\theta_\beta-\cos\theta_\alpha)^2\,d\theta_1\ldots d\theta_\nu = \frac{\pi}{2^{(\nu-1)^2}}. \qquad (3.7.9)$$

PROOF. From the formula

$$\sin^2(\theta_\beta-\theta_\alpha)=\frac{1}{4}\left|e^{2i\theta_\alpha}-e^{2i\theta_\beta}\right|^2$$

and from the orthogonality of the functions $e^{im\theta}$ it follows that

$$\int_{\pi>\theta_1>\cdots>\theta_\nu>0}\cdots\int \prod_{1\leqslant\alpha<\beta\leqslant\nu}\sin^2(\theta_\beta-\theta_\alpha)\,d\theta_1\ldots d\theta_\nu$$

$$=\frac{1}{\nu!\,2^\nu\cdot 2^{\nu(\nu-1)}}\int_0^{2\pi}\cdots\int_0^{2\pi}\prod_{1\leqslant\alpha<\beta\leqslant\nu}\left|e^{i\theta_\alpha}-e^{i\theta_\beta}\right|^2 d\theta_1\ldots d\theta_\nu=\frac{(2\pi)^\nu\cdot\nu!}{\nu!\,2^{\nu^2}}=\frac{(2\pi)^\nu}{2^{\nu^2}}.$$

From the orthogonality of the functions $\cos m\theta$ we have

$$\int_{\pi>\theta_1>\cdots>\theta_\nu>0}\cdots\int \prod_{1\leqslant\alpha<\beta\leqslant\nu}(\cos\theta_\beta-\cos\theta_\alpha)^2\,d\theta_1\ldots d\theta_\nu$$

$$=\frac{1}{2^\nu\cdot\nu!}\int_0^{2\pi}\cdots\int_0^{2\pi}\left\{\det(\cos^{i-1}\theta_j)_1^\nu\right\}^2 d\theta_1\ldots d\theta_\nu$$

$$=\frac{1}{2^\nu\cdot\nu!\,2^{(\nu-1)(\nu-2)}}\int_0^{2\pi}\cdots\int_0^{2\pi}\left\{\det(\cos(i-1)\theta_j)_1^\nu\right\}^2 d\theta_1\ldots d\theta_\nu$$

$$=\frac{1}{2^\nu\nu!\,2^{(\nu-1)(\nu-2)}}\cdot\nu!\cdot 2\cdot\pi^\nu=\frac{\pi^\nu}{2^{(\nu-1)^2}},$$

since $\cos^2\theta=\frac{1}{2}(\cos 2\theta+1)$ and $\cos^m\theta=2^{-(m-1)}\cos m\theta+\cdots$.

Now we consider the set of orthogonal matrices G of the form

$$G=\Gamma M\Gamma', \tag{3.7.10}$$

where Γ is a matrix in O_n^+, and

$$M=\begin{pmatrix}\cos\theta_1 & \sin\theta_1\\ -\sin\theta_1 & \cos\theta_1\end{pmatrix}\dot{+}\begin{pmatrix}\cos\theta_2 & \sin\theta_2\\ -\sin\theta_2 & \cos\theta_2\end{pmatrix}\dot{+}\cdots,\quad \pi\geqslant\theta_1\geqslant\cdots\geqslant\theta_\nu\geqslant 0. \tag{3.7.11}$$

For $n=2\nu$ the last term of this sum is $\begin{pmatrix}\cos\theta_\nu & \sin\theta_\nu\\ -\sin\theta_\nu & \cos\theta_\nu\end{pmatrix}$, and for $n=2\nu+1$ the last terms are $\begin{pmatrix}\cos\theta_\nu & \sin\theta_\nu\\ -\sin\theta_\nu & \cos\theta_\nu\end{pmatrix}\dot{+}1$. The set of all such matrices will again be denoted by G, and we denote the volume of this set by $V(G)$. For odd n the set G coincides with O_n^+, and consequently,

$$V(G) = (8\pi)^{\frac{n(n-1)}{4}} \prod_{\alpha=2}^{n} \frac{\Gamma\left(\frac{\alpha-1}{2}\right)}{\Gamma(\alpha-1)} \tag{3.7.12}$$

using (3.7.7). For even n, any matrix in O_n^+ can again be represented in the form (3.7.10), but only if the determinant of Γ is allowed to be equal to both $+1$ and -1. It follows that

$$V(G) = \frac{1}{2}(8\pi)^{\frac{n(n-1)}{4}} \prod_{\alpha=2}^{n} \frac{\Gamma\left(\frac{\alpha-1}{2}\right)}{\Gamma(\alpha-1)}. \tag{3.7.13}$$

Now we compute $V(G)$ in another way. From the relations

$$G = \Gamma M \Gamma' = \Gamma_1 M_1 \Gamma_1', \qquad \pi > \theta_1 > \ldots > \theta_\nu > 0,$$

we deduce $\Gamma = \Gamma_1 \Delta$, $M = M_1$, where

$$\Delta = \begin{pmatrix} \cos\delta_1 & \sin\delta_1 \\ -\sin\delta_1 & \cos\delta_1 \end{pmatrix} \dotplus \begin{pmatrix} \cos\delta_2 & \sin\delta_2 \\ -\sin\delta_2 & \cos\delta_2 \end{pmatrix} \dotplus \cdots.$$

So it is possible to establish a one-to-one correspondence between the elements of the sets G and $\Sigma \times M$ (where, as above, $\Sigma = \Gamma/\Delta$), with the exclusion, of course, of a manifold of lower dimension.

Using the method applied before in this chapter we find that the volume element of G is equal to

$$\dot{G} = b_n \prod_{1 \leqslant \alpha < \beta \leqslant \nu} (\cos\theta_\alpha - \cos\theta_\beta)^2 \prod_{\alpha=1}^{\nu} d\theta_\alpha \cdot \dot{\Sigma},$$

where

$$b_n = \begin{cases} 2^{2\nu(\nu-1)+\frac{\nu}{2}}, & n = 2\nu, \\ 2^{2\nu^2+\frac{\nu}{2}}, & n = 2\nu+1. \end{cases}$$

We set

$$V(\Sigma) = \int_{\Sigma} \dot{\Sigma}.$$

Then

$$V(G) = b_n V(\Sigma) \int_{\pi > \theta_1 > \ldots > \theta_\nu > 0} \cdots \int \prod_{1 \leqslant \alpha < \beta \leqslant \nu} (\cos\theta_\alpha - \cos\theta_\beta)^2 \, d\theta_1 \ldots d\theta_\nu$$

$$= b_n V(\Sigma) \cdot \frac{\pi^\nu}{2^{(\nu-1)^2}}. \tag{3.7.14}$$

From Theorem 3.6.1 we know that

$$V(\mathfrak{K}) = \varkappa_n \cdot V(\Sigma) \cdot \int_{\pi > \theta_1 > \cdots > \theta_\nu > 0} \cdots \int \prod_{1 \leqslant \alpha < \beta \leqslant \nu} \sin^2(\theta_\alpha - \theta_\beta)\, d\theta_1 \ldots d\theta_\nu,$$

since by (3.7.8) we have

$$V(\mathfrak{K}) = \varkappa_n V(\Sigma) \cdot \frac{(2\pi)^\nu}{2^{\nu^2}}. \tag{3.7.15}$$

Combining (3.7.14) and (3.7.15), we have

$$V(\mathfrak{K}) = \frac{\varkappa_n}{b_n} \cdot \frac{2^{(\nu-1)2}}{\pi^\nu} \cdot \frac{(2\pi)^\nu}{2^{\nu^2}} \cdot V(G).$$

Substituting the values of κ_n, b_n and $V(G)$ we finally obtain

$$V(\mathfrak{K}) = \lambda_n \frac{1}{2^{\nu-1}} (8\pi)^{\frac{n(n-1)}{4}} \cdot \prod_{\alpha=1}^{n-1} \frac{\Gamma\left(\frac{\alpha}{2}\right)}{\Gamma(\alpha)}, \qquad \lambda_n = \begin{cases} 1, & n = 2\nu, \\ 2^{-\nu}, & n = 2\nu + 1. \end{cases}$$

For even n the set $\mathfrak{K}$ coincides with $\mathfrak{C}_{\mathrm{III}}$, the characteristic manifold of the domain $\mathfrak{R}_{\mathrm{III}}$. In this way we have, in particular, obtained the volume of $\mathfrak{C}_{\mathrm{III}}$.

Chapter IV

SOME GENERAL THEOREMS AND THEIR APPLICATIONS

4.1. **Introduction.** Let $\mathfrak{R}$ be a bounded domain in the $2n$-dimensional Euclidean space of n complex variables $z=(z_1, z_2, \cdots, z_n)$. It is well known that a function $f(z)$, which is analytic in $\mathfrak{R}$, cannot attain its maximum modulus at an interior point of $\mathfrak{R}$. Let $\mathfrak{C}$ be a manifold on the boundary of $\mathfrak{R}$, with the following properties:

1. Every function which is analytic in $\mathfrak{R}$ attains its maximum modulus on the manifold $\mathfrak{C}$.

2. For any point a of $\mathfrak{C}$ there exists a function $f(z)$, analytic in $\mathfrak{R}$, which attains its maximum modulus at a point a.

Such a manifold we shall call the characteristic manifold of the domain $\mathfrak{R}$. It is evident that $\mathfrak{C}$ is uniquely determined by $\mathfrak{R}$. It can be readily shown that $\mathfrak{C}$ is closed and that any function which is analytic in the neighborhood of every point of $\mathfrak{C}$ is uniquely determined by its values on $\mathfrak{C}$. Hence the real dimension n_1 of the manifold $\mathfrak{C}$ is not less than n. We shall denote by $\xi=(\xi_1, \cdots, \xi_{n_1})$ the variables on $\mathfrak{C}$, by $d\xi d\bar{\xi}'=\sum_{i=1}^{n_1}|d\xi_i|^2$ the metric on $\mathfrak{C}$, and by $\dot{\xi}$ the volume element on $\mathfrak{C}$.

If the domain $\mathfrak{R}$ admits of analytic mappings which carry it into itself, then, of course, these mappings will also carry $\mathfrak{C}$ into itself. In particular, $\mathfrak{C}$ is invariant with respect to the group of motions $\mathfrak{R}$ which leaves a given point unchanged. If the domain $\mathfrak{R}$ admits a transformation group $z=e^{i\theta}w$, we shall call $\mathfrak{R}$ a circular domain; if, in addition to z, $\mathfrak{R}$ also contains the point rz, $0\leqq r\leqq 1$, then we shall call $\mathfrak{R}$ a complete circular domain. It is evident that if $\mathfrak{R}$ is a circular domain, then $\mathfrak{C}$ is also.

Suppose now $\mathfrak{R}$ is a circular domain. Let us consider the vector $z^{[f]}$ with components

$$\sqrt{\frac{f!}{a_1!\ldots a_n!}}\, z_1^{a_1}\ldots z_n^{a_n}, \qquad a_1+\ldots+a_n=f. \tag{4.1.1}$$

The dimension of the vector $z^{[f]}$ is

$$N_f=\frac{1}{f!}\,n(n+1)\ldots(n+f-1)=\binom{n+f-1}{f}.$$

For $f \neq g$

$$\int_{\mathfrak{R}} (\overline{z^{[f]}})' \cdot z^{[g]} \cdot \dot{z} = 0$$

and

$$\int_{\mathfrak{C}} (\overline{\xi^{[f]}})' \cdot \xi^{[g]} \cdot \dot{\xi} = 0.$$

In fact, since $\mathfrak{R}$ is a circular domain, by the change of variables $z = e^{i\theta} w$, we obtain

$$\int_{\mathfrak{R}} (\overline{z^{[f]}})' \cdot z^{[g]} \cdot \dot{z} = e^{i\theta(f-g)} \int_{\mathfrak{R}} (\overline{w^{[f]}})' \cdot w^{[g]} \cdot \dot{w},$$

from which our assertion follows, since $e^{i\theta(f-g)} \neq 1$.

We shall set

$$\int_{\mathfrak{R}} (\overline{z^{[f]}})' \cdot z^{[f]} \cdot \dot{z} = H_1 \tag{4.1.2}$$

and

$$\int_{\mathfrak{C}} (\overline{\xi^{[f]}})' \cdot \xi^{[f]} \cdot \dot{\xi} = H_2. \tag{4.1.3}$$

It is clear that H_1 and H_2 are positive-definite Hermitian matrices of order N_f. There exists a matrix Γ such that

$$\overline{\Gamma}' H_1 \Gamma = \Lambda, \qquad \overline{\Gamma}' H_2 \Gamma = I,$$

where $\Lambda = [\beta_1^f, \beta_2^f, \cdots, \beta_{N_f}^f]$ is the diagonal matrix.

We shall set

$$z_f = z^{[f]} \cdot \Gamma, \qquad \xi_f = \xi^{[f]} \cdot \Gamma,$$

and we shall denote by $\{\varphi_\nu^f(z)\}$ the components of the vector z_f. Then we have

$$\int_{\mathfrak{R}} \varphi_\nu^f(z) \overline{\varphi_\mu^g(z)} \dot{z} = \delta_{\nu\mu} \cdot \delta_{fg} \cdot \beta_\nu^f \tag{4.1.4}$$

and

$$\int_{\mathfrak{C}} \varphi_\nu^f(\xi) \overline{\varphi_\mu^g(\xi)} \dot{\xi} = \delta_{\nu\mu} \cdot \delta_{fg}. \tag{4.1.5}$$

Thus the functions $\{(\beta_\nu^f)^{-1/2} \varphi_\nu^f(z)\}$ form an orthonormal system in the domain $\mathfrak{R}$. The following theorem is well known (H. Cartan [**1**]).

THEOREM 4.1.1. *For a complete circular domain* $\mathfrak{R}$ *the system of functions*

$$\left(\beta_\nu^f\right)^{-\frac{1}{2}} \varphi_\nu^f(z), \quad \begin{matrix} f=0,\ 1,\ 2,\ \ldots, \\ \nu=1,\ 2,\ \ldots,\ N_f, \end{matrix} \tag{4.1.6}$$

is a complete orthonormal system in the domain $\mathfrak{R}$. *On the other hand, the system* $\{\varphi_\nu^f(\xi)\}$ *is orthonormal, but in general not complete in the space of functions which are continuous on* $\mathfrak{C}$.

It is also well known that the series

$$\sum_{f=0}^{\infty} \sum_{\nu=1}^{N_f} \frac{\varphi_\nu^f(z)\overline{\varphi_\nu^f(w)}}{\beta_\nu^f} = K(z, \overline{w})$$

converges uniformly for any z and w which lie in the interior of $\mathfrak{R}$, representing there a function called the Bergman kernel.[12]

The sum of the series (if it converges)

$$\sum_{f=0}^{\infty} \sum_{\nu=0}^{N_f} \varphi_\nu^f(z)\overline{\varphi_\nu^f(\xi)} = H(z, \overline{\xi})$$

we shall call the Cauchy kernel for the domain $\mathfrak{R}$.

Finally, we shall call the function

$$P(z, \xi) = \frac{|H(z, \overline{\xi})|^2}{H(z, \overline{z})}$$

the Poisson kernel for the domain $\mathfrak{R}$.

This chapter deals with the direct methods of determination of these kernels.

4.2. **The Bergman kernel.** Let $\mathfrak{R}$ be a bounded domain which contains the origin, Γ a group of analytic mappings of $\mathfrak{R}$ onto itself, and Γ_0 a subgroup of Γ which leaves the origin fixed. It is well known (H. Cartan [1]) that an element of Γ_0 is fully determined by its linear terms in the neighborhood of the origin, i.e., the mapping of $\mathfrak{R}$ onto itself which has the form

$$w_i = \sum_{j=1}^{n} u_{ij} z_j + \sum_{\substack{m_1 \ldots m_n \\ m_1 + \ldots + m_n \geqslant 2}} a_{m_1 \ldots m_n}^{(i)} z_1^{m_1} \ldots z_n^{m_n}, \tag{4.2.1}$$

[12] Translator's note. In the Russian literature the Bergman kernel is usually called the "kernel function of the domain". In this book such a designation would not be very appropriate since we are dealing with three kernels to which this designation could apply.

is fully determined if the matrix $(u_{ij})_1^n$ is given. As it is well known that Γ_0 is compact, it can be assumed without loss of generality that the matrices $(u_{ij})=U$ which form the representation of Γ_0 are unitary. The letter U we shall also use for denoting the nonlinear transformation (4.2.1) itself, determined by the linear element U.

Let us now consider the set of cosets of Γ/Γ_0. All group transformations belonging to one and the same coset carry into the origin one and the same point a. The totality of all such points a forms in $\mathfrak{R}$ some set $\mathfrak{M}$. It is called a transitive set with respect to the group Γ which contains the origin. Thus any element of Γ is uniquely determined by a point a of $\mathfrak{M}$, and by the unitary matrix U of Γ_0. We shall write the transformations determined by the elements of Γ in the form

$$w=f(z, a, U), \quad a\in\mathfrak{M}, \quad U\in\Gamma_0. \tag{4.2.2}$$

Suppose

$$z=f(x, b, V), \quad b\in\mathfrak{M}, \quad V\in\Gamma_0, \tag{4.2.3}$$

is another transformation, and

$$w=f(f(x, b, V), a, U)=f(x, c, W) \tag{4.2.4}$$

is the product of the transformations (4.2.2) and (4.2.3). Setting $w=0$, we at once obtain

$$a=f(c, b, V). \tag{4.2.5}$$

Differentiation of (4.2.4) yields

$$\frac{\partial f_i(x, c, W)}{\partial x_j}=\sum_{k=1}^{n}\frac{\partial f_i(z, a, U)}{\partial z_k}\,\frac{\partial f_k(x, b, V)}{\partial x_j}. \tag{4.2.6}$$

We shall denote the Jacobian of the transformation (4.2.2) by

$$J(z, a, U)=(a_{ij}), \quad a_{ij}=\frac{\partial f_i(z, a, U)}{\partial z_j}.$$

By setting $x=c$ in (4.2.6) we obtain $z=a$. Hence

$$J(c, c, W)=J(a, a, U)\cdot J(c, b, V).$$

By a change of notation we obtain

$$J(x, x, W)=J(z, z, U)\,J(x, b, V). \tag{4.2.7}$$

This formula is valid for x and z of $\mathfrak{M}$ which satisfy the relation

$$z=f(x, b, V). \tag{4.2.8}$$

If we have another transformation

$$u=f(x,b,V_0),$$

then the mapping of u into z leaves the origin unchanged. Hence

$$U_0=\left(\frac{\partial z}{\partial u}\right)_{z=0}$$

is a unitary matrix. Whence follows that

$$\{J(x,b,V)\}_{x=b}=\left(\frac{\partial z}{\partial u}\right)_{z=0}\cdot\{J(x,b,V_0)\}_{x=b},$$

so that we have

$$J(b,b,V)=U_0 J(b,b,V_0), \tag{4.2.9}$$

where U_0 is the unitary matrix of Γ_0. Thus

$$\overline{J(z,z,V)}'\cdot J(z,z,V)=\overline{J(z,z,V_0)}'\cdot J(z,z,V_0). \tag{4.2.10}$$

This shows that $\bar{J}'J$ depends on the coset of Γ/Γ_0 but does not depend on the choice of representative of this coset. Hence we can write

$$|\det J(z,z,V)|^2=Q(z,\bar{z}).$$

From (4.2.7) we obtain for z and x, lying in $\mathfrak{M}$ and satisfying the relation (4.2.8), the formula

$$Q(x,\bar{x})=Q(z,\bar{z})|\det J(x,b,V)|^2. \tag{4.2.11}$$

Bergman [1] has proved that under the transformation (4.2.8) the Bergman kernel of the domain $\mathfrak{R}$ changes according to the law

$$K(x,\bar{x})=K(z,\bar{z})|\det J(x,b,V)|^2. \tag{4.2.12}$$

Thus for z and x of $\mathfrak{M}$

$$\frac{K(x,\bar{x})}{Q(x,\bar{x})}=\frac{K(z,\bar{z})}{Q(z,\bar{z})}. \tag{4.2.13}$$

THEOREM 4.2.1. *If $\mathfrak{R}$ is a bounded circular domain, then for z lying in $\mathfrak{M}$, we have*

$$K(z,\bar{z})=\frac{1}{\Omega}Q(z,\bar{z}),$$

where Ω is the complete volume of $\mathfrak{R}$.

PROOF. In view of §4.1, we can propose the following process of constructing an orthonormal system of functions.

We orthonormalize the terms

$$z_1^{a_1} \cdot z_2^{a_2} \cdot \ldots \cdot z_n^{a_n}, \qquad a_1 + \ldots + a_n = m,$$

for a given m, and we take the totality of all such functions for $m=0,1,2,\cdots$. This totality forms a complete orthonormal system.

Among the functions $\varphi_\nu(z)$ obtained by this process, we have the constant $\Omega^{-1/2}$, whereas the other functions are homogeneous forms of order $m \geqq 1$. Hence

$$\varphi_0(z) = \Omega^{-\frac{1}{2}}, \qquad \varphi_\nu(0) = 0, \qquad \nu \geqq 1.$$

Therefore from equation

$$K(z, \bar{z}) = \sum_{\nu=0}^{\infty} \varphi_\nu(z) \overline{\varphi_\nu(z)}$$

we obtain at once

$$K(0, 0) = \frac{1}{\Omega}.$$

On the other hand, by the definition of $Q(z, \bar{z})$ we have $Q(0,0)=1$. Hence the theorem follows from (4.2.13).

Assuming now that $\mathfrak{R}$ is a transitive domain (i.e., $\mathfrak{R} \equiv \mathfrak{M}$), let us ascertain the geometrical properties of $Q(z, \bar{z})$. From (4.2.7) we have

$$J(x, x, W) dx' = J(z, z, U) dz',$$

hence

$$\overline{dx} \cdot \overline{J(x, x, W)'} \cdot J(x, x, W) \cdot dx' = \overline{dz} \cdot \overline{J(z, z, U)'} \cdot J(z, z, U) \cdot dz'.$$

This invariant form can be considered as the metric of our space. The volume element in this metric is

$$|\det J(z, z, U)|^2 \dot{z} = Q(z, \bar{z}) \cdot \dot{z},$$

so that $Q(z, \bar{z})$ can be called the *volume density*.

From Theorem 4.2.1 we obtain the following proposition:

The Bergman kernel for any transitive circular region is equal to the ratio of the volume density to the Euclidean volume of the domain.

In the subsequent sections we shall determine the Bergman kernel for our four types of classical domains on the basis of the above considerations only, without the use of complete orthonormal systems.[13]

4.3. **Bergman kernels for the domains $\mathfrak{R}_{\rm I}$, $\mathfrak{R}_{\rm II}$ and $\mathfrak{R}_{\rm III}$.**

1°. The group Γ for the domain $\mathfrak{R}_{\rm I}$ consists of the following transformations (see L. K. Hua [1]):

$$Z_1=(AZ+B)(CZ+D)^{-1}, \tag{4.3.1}$$

where A, B, C, D are matrices of dimensions $m\times m$, $m\times n$, $n\times m$ and $n\times n$, respectively, satisfying the relations

$$\bar{A}A'-\bar{B}B'=I^{(m)},\quad \bar{A}C'=\bar{B}D',\quad \bar{C}C'-\bar{D}D'=-I^{(n)}.$$

For $m=n$, we assume, moreover, that

$$\det\begin{pmatrix} A & B \\ C & D\end{pmatrix}=+1. \tag{4.3.2}$$

Let us find the transformations which carry an arbitrary point $Z=P$ into the origin. By the definition of the domain $\mathfrak{R}_{\rm I}$ we have

$$I^{(m)}-\overline{P}P'>0.$$

Thus it follows from Theorem 2.1.2 that also

$$I^{(n)}-P'\overline{P}>0.$$

It is known that there exists an $m\times m$ matrix Q and $n\times n$ matrix R such that

$$\overline{Q}\left(I^{(m)}-\overline{P}P'\right)Q'=I^{(m)},\quad \overline{R}\left(I^{(n)}-P'\overline{P}\right)R'=I^{(n)}. \tag{4.3.3}$$

The transformation

$$Z_1=Q(Z-P)(I^{(n)}-\overline{P'}Z)^{-1}R^{-1} \tag{4.3.4}$$

carries P into the origin. It is easy to see that this transformation is of the form (4.3.1).

Differentiating (4.3.4), we obtain

$$dZ_1=Q\left\{dZ\cdot(I-\bar{P}'Z)^{-1}+(Z-P)\,d\,(I-\bar{P}'Z)^{-1}\right\}R^{-1}.$$

We shall set $Z=P$. Then

$$dZ_1=Q\cdot dZ\cdot(I-\bar{P}'P)^{-1}R^{-1}=Q\cdot dZ\cdot\bar{R}',$$

i.e., at the point $Z=P$

$$\dot{Z}_1=\left|(\det Q)^m\cdot(\det\bar{R}')^n\right|^2\cdot\dot{Z}=\{\det(I-P\bar{P}')\}^{-m-n}\cdot\dot{Z}.$$

[13] Translator's note. The 4 domains in question, $\mathfrak{R}_{\rm I}$, $\mathfrak{R}_{\rm II}$, $\mathfrak{R}_{\rm III}$ and $\mathfrak{R}_{\rm IV}$, are defined by the author in the introduction to the book.

Hence

$$Q(Z, \bar{Z}) = \{\det(I - Z\bar{Z}')\}^{-(m+n)}.$$

Using the results of §4.2, we obtain the following theorem.

THEOREM 4.3.1. *The Bergman kernel of the domain* $\mathfrak{R}_{\mathrm{I}}$ *is*

$$\frac{1}{V(\mathfrak{R}_{\mathrm{I}})} \cdot \{\det(I - Z\bar{Z}')\}^{-(m+n)}, \tag{4.3.5}$$

where, by (2.2.2)

$$V(\mathfrak{R}_{\mathrm{I}}) = \frac{1!\,2!\ldots(m-1)!\,1!\,2!\ldots(n-1)!}{1!\,2!\ldots(m+n-1)!}\pi^{mn}.$$

2°. The group Γ of the domain $\mathfrak{R}_{\mathrm{II}}$ consists of transformations of the form

$$Z_1 = (AZ + B)(\bar{B}Z + \bar{A})^{-1}, \tag{4.3.6}$$

where

$$A'B = B'A, \quad \bar{A}A' - \bar{B}B' = I.$$

Suppose P is a point of $\mathfrak{R}_{\mathrm{II}}$. A matrix R can be found such that

$$\bar{R}(I - \bar{P}P')R' = I. \tag{4.3.7}$$

The transformation

$$Z_1 = R(Z - P)(I - \bar{P}Z)^{-1}\bar{R}^{-1}, \tag{4.3.8}$$

belonging to Γ, carries the point P into the origin.

Differentiation of (4.3.8) yields

$$dZ_1 = R\{dZ \cdot (I - \bar{P}Z)^{-1} + (Z - P)\,d(I - \bar{P}Z)^{-1}\}\bar{R}^{-1}.$$

Setting $Z = P$, we obtain

$$dZ_1 = R \cdot dZ \cdot (I - \bar{P}P)^{-1}\bar{R}^{-1} = R \cdot dZ \cdot R'.$$

Hence at the point $Z = P$

$$\dot{Z}_1 = |(\det R)^{n+1}|^2 \cdot \dot{Z} = \{\det(I - P\bar{P})\}^{-(n+1)} \cdot \dot{Z}.$$

Thus

$$Q(Z, \bar{Z}) = \{\det(I - Z\bar{Z})\}^{-(n+1)}.$$

THEOREM 4.3.2. *The Bergman kernel of the domain* $\mathfrak{R}_{\mathrm{II}}$ *is*

$$\frac{1}{V(\mathfrak{R}_{\mathrm{II}})} \cdot \{\det(I - Z\bar{Z})\}^{-(n+1)}, \tag{4.3.9}$$

where, by (2.3.2),

$$V(\mathfrak{R}_{II}) = \pi^{\frac{n(n+1)}{2}} \cdot \frac{2!\,4!\ldots(2n-2)!}{n!\,(n+1)!\ldots(2n-1)!}.$$

3°. The group Γ of the domain $\mathfrak{R}_{III}$ consists of transformations of the form

$$Z_1 = (AZ + B)(-\bar{B}Z + \bar{A})^{-1}, \tag{4.3.10}$$

where

$$A'B = -B'A, \quad \bar{A}'A - \bar{B}'B = I.$$

Suppose P is a point of $\mathfrak{R}_{III}$, i.e., $I + P\overline{P} > 0$. Then a matrix Q can be found such that

$$\bar{Q}(I + P\bar{P})Q' = I.$$

Then in Γ we have the transformation

$$Z_1 = Q(Z - P)(I + \bar{P}Z)^{-1}\bar{Q}^{-1}, \tag{4.3.11}$$

which carries the point P into the origin.

By differentiation of (4.3.11) we have

$$dZ_1 = Q\{dZ \cdot (I + \bar{P}Z)^{-1} + (Z - P)\,d\,(I + \bar{P}Z)^{-1}\}\,\bar{Q}^{-1}.$$

For $Z = P$, we obtain

$$dZ_1 = Q \cdot dZ \cdot (I + \bar{P}P)^{-1} \cdot \bar{Q}^{-1} = Q \cdot dZ \cdot Q'.$$

Hence at the point $Z = P$

$$\dot{Z}_1 = |(\det Q)^{n-1}|^2\,\dot{Z} = \{\det(I + \bar{P}P)\}^{-n+1} \cdot \dot{Z}.$$

Therefore

$$Q(Z, \bar{Z}) = \{\det(I + Z\bar{Z})\}^{-n+1}.$$

THEOREM 4.3.3. *The Bergman kernel of the domain* $\mathfrak{R}_{III}$ *is*

$$\frac{1}{V(\mathfrak{R}_{III})}\{\det(I + Z\bar{Z})\}^{-n+1},$$

where, by (2.4.2),

$$V(\mathfrak{R}_{III}) = \pi^{\frac{n(n-1)}{2}} \cdot \frac{2!\,4!\ldots(2n-4)!}{(n-1)!\,n!\ldots(2n-3)!}.$$

4.4. **The Bergman kernel for the domain** $\mathfrak{R}_{IV}$. The group Γ of the domain $\mathfrak{R}_{IV}$ consists of transformations of the form

$$w=\left\{\left[\left(\frac{1}{2}(zz'+1),\ \frac{i}{2}(zz'-1)\right)A'+zB'\right]\binom{1}{i}\right\}^{-1}$$
$$\times\left\{\left(\frac{1}{2}(zz'+1),\ \frac{i}{2}(zz'-1)\right)C'+zD'\right\}, \tag{4.4.1}$$

where A, B, C and D are real matrices of dimensions 2×2, $2\times n$, $n\times 2$ and $n\times n$, respectively, satisfying the relations

$$\begin{pmatrix} A & B \\ C & D\end{pmatrix}\begin{pmatrix} I^{(2)} & 0 \\ 0 & -I^{(n)}\end{pmatrix}\begin{pmatrix} A & B \\ C & D\end{pmatrix}'=\begin{pmatrix} I^{(2)} & 0 \\ 0 & -I^{(n)}\end{pmatrix} \tag{4.4.2}$$

and

$$\det\begin{pmatrix} A & B \\ C & D\end{pmatrix}=+1. \tag{4.4.3}$$

We shall now find the transformations of Γ which carry the point z_0 into the origin. Proceeding from the vector z_0 we shall construct the $2\times n$ matrix X_0 as follows:

$$X_0=2\begin{pmatrix} z_0z_0'+1 & i\,(z_0z_0'-1) \\ \overline{z_0z_0'}+1 & -i\,(\overline{z_0z_0'}-1)\end{pmatrix}^{-1}\begin{pmatrix} z_0 \\ \bar{z}_0\end{pmatrix}=2A_0^{-1}\begin{pmatrix} z_0 \\ \bar{z}_0\end{pmatrix}$$
$$=\frac{1}{1-|z_0z_0'|^2}\begin{pmatrix} z_0+\bar{z}_0-(\overline{z_0z_0'}\cdot z_0+z_0z_0'\cdot\bar{z}_0) \\ i\,(z_0-\bar{z}_0)+i\,(\overline{z_0z_0'}\cdot z_0-z_0z_0'\cdot\bar{z}_0)\end{pmatrix}. \tag{4.4.4}$$

This matrix is evidently real. We have

$$I-X_0X_0'=\bar{A}_0^{-1}\left(\bar{A}_0A_0'-4\,\overline{\begin{pmatrix} z_0 \\ \bar{z}_0\end{pmatrix}}\begin{pmatrix} z_0 \\ \bar{z}_0\end{pmatrix}'\right)\cdot A_0'^{-1}$$
$$=2\bar{A}_0^{-1}\begin{pmatrix} 1+|z_0z_0'|^2-2\bar{z}_0z_0' & 0 \\ 0 & 1+|z_0z_0'|^2-2\bar{z}_0z_0'\end{pmatrix}\cdot A_0'^{-1}. \tag{4.4.5}$$

Hence

$$(I-X_0X_0')^{-1}=\frac{1}{2\,(1+|z_0z_0'|^2-2\bar{z}_0z_0')}\,A_0'\bar{A}_0=\frac{1}{1+|z_0z_0'|^2-2\bar{z}_0z_0'}$$
$$\times\begin{pmatrix} (z_0z_0'+1)(\overline{z_0z_0'}+1) & i\,(z_0z_0'-\overline{z_0z_0'}) \\ i\,(z_0z_0'-\overline{z_0z_0'}) & (z_0z_0'-1)(\overline{z_0z_0'}-1)\end{pmatrix}=A'A, \tag{4.4.6}$$

where

$$A=\frac{1}{2}(1+|z_0z_0'|^2-2\bar{z}_0z_0')^{-\frac{1}{2}}\begin{pmatrix} -i\,(z_0z_0'-\overline{z_0z_0'}) & z_0z_0'+\overline{z_0z_0'}-2 \\ z_0z_0'+\overline{z_0z_0'}+2 & i\,(z_0z_0'-\overline{z_0z_0'}) \end{pmatrix}. \tag{4.4.7}$$

We shall choose a D which satisfies the condition $D(I^{(n)}-X_0'X_0)D'=I^{(n)}$; then the transformation

$$w=\left\{\left[\left(\frac{1}{2}(zz'+1),\ \frac{i}{2}(zz'-1)\right)A'-zX_0'A'\right]\left(\begin{matrix}1\\ i\end{matrix}\right)\right\}^{-1} \times\left\{zD'-\left(\frac{1}{2}(zz'+1),\ \frac{i}{2}(zz'-1)\right)X_0D'\right\} \tag{4.4.8}$$

has the form (4.4.1) and carries the point z_0 into the origin.

In addition,

$$\det A=\det D=\frac{1-|z_0z_0'|^2}{1+|z_0z_0'|^2-2\bar{z}_0z_0'}. \tag{4.4.9}$$

Differentiation of (4.4.8) yields ($z=(z^{(1)},\cdots,z^{(n)})$)

$$\begin{aligned} dw=&\left\{dz\cdot D'-\left(\sum_{p=1}^n z^{(p)}dz^{(p)},\ i\sum_{p=1}^n z^{(p)}dz^{(p)}\right)X_0D'\right\}\\ &\times\left\{\left[\left(\frac{1}{2}(zz'+1),\ \frac{i}{2}(zz'-1)\right)A'-zX_0'A'\right]\left(\begin{matrix}1\\ i\end{matrix}\right)\right\}^{-1}\\ &+\left\{zD'-\left(\frac{1}{2}(zz'+1),\ \frac{i}{2}(zz'-1)\right)X_0D'\right\}\\ &\times d\left\{\left[\left(\frac{1}{2}(zz'+1),\ \frac{i}{2}(zz'-1)\right)A'-zX_0'A'\right]\left(\begin{matrix}1\\ i\end{matrix}\right)\right\}^{-1}. \end{aligned}$$

Setting $z=z_0$, we obtain

$$\begin{aligned} dw=&\left\{dz\cdot D'-\left(\sum_{p=1}^n z_0^{(p)}dz^{(p)},\ i\sum_{p=1}^n z_0^{(p)}dz^{(p)}\right)X_0D'\right\}\\ &\times\left\{\left[\left(\frac{1}{2}(z_0z_0'+1),\ \frac{i}{2}(z_0z_0'-1)\right)A'-z_0X_0'A'\right]\left(\begin{matrix}1\\ i\end{matrix}\right)\right\}^{-1}, \end{aligned}$$

i.e.,

$$\begin{aligned} dw=&\{dz\cdot D'-d(zz')\cdot(1,i)X_0D'\}\\ &\times\left\{\left[\left(\frac{1}{2}(z_0z_0'+1),\ \frac{i}{2}(z_0z_0'-1)\right)A'-z_0X_0'A'\right]\left(\begin{matrix}1\\ i\end{matrix}\right)\right\}^{-1} \end{aligned} \tag{4.4.10}$$

Using (4.4.4) and (4.4.7), we have

$$dw=-i\,dz\cdot\left\{I-2\frac{z_0'\bar{z}_0-\overline{z_0z_0'}\cdot z_0'z_0}{1-|z_0z_0'|^2}\right\}\cdot D'\cdot(1+|z_0z_0'|^2-2\bar{z}_0z_0')^{-\frac{1}{2}}. \tag{4.4.11}$$

From (4.4.9) we obtain

$$\det\left(\frac{\partial w}{\partial z}\right)_{z=z_0}$$

$$= \det\left\{I - 2\frac{z_0'\bar{z}_0 - \overline{z_0 z_0'} \cdot z_0' z_0}{1 - |z_0 z_0'|^2}\right\} \cdot \frac{1 - |z_0 z_0'|^2}{(1 + |z_0 z_0'|^2 - 2\bar{z}_0 z_0')}$$

$$\times (1 + |z_0 z_0'|^2 - 2\bar{z}_0 z_0')^{-\frac{n}{2}},$$

or

$$\left|\det\left(\frac{\partial w}{\partial z}\right)_{z=z_0}\right|^2 = (1 + |z_0 z_0'|^2 - 2\bar{z}_0 z_0')^{-n}. \tag{4.4.12}$$

Here the identity

$$\det\left\{I - 2\frac{z_0'\bar{z}_0 - \overline{z_0 z_0'} \cdot z_0' z_0}{1 - |z_0 z_0'|^2}\right\} = \frac{1 + |z_0 z_0'|^2 - 2\bar{z}_0 z_0'}{1 - |z_0 z_0'|^2} \tag{4.4.13}$$

is used, which follows from the relation $\det(I - u'\bar{v}) = 1 - \bar{v}u'$ (see Theorem 2.1.2).

Thus we arrived at the theorem:

THEOREM 4.4.1. *The Bergman kernel of the domain* $\mathfrak{R}_{IV}$ *is*

$$\frac{1}{V(\mathfrak{R}_{IV})}(1 + |zz'|^2 - 2\bar{z}z')^{-n},$$

where, by (2.5.7),

$$V(\mathfrak{R}_{IV}) = \frac{\pi^n}{2^{n-1} \cdot n!}.$$

4.5. **The Cauchy kernel.** Let us now pass to the study of the Cauchy kernel

$$\sum_{f=0}^{\infty} \sum_{\nu=1}^{N_f} \varphi_\nu^f(z)\overline{\varphi_\nu^f(\xi)} = H(z, \bar{\xi}). \tag{4.5.1}$$

(Here z belongs to $\mathfrak{R}$, and ξ to $\mathfrak{C}$.)

Suppose Γ_0 is a group of motions of $\mathfrak{R}$ which leave the origin unchanged. We shall assume that $\mathfrak{C}$ is transitive with respect to Γ_0, i.e., that any two points of $\mathfrak{C}$ can be carried into each other by a transformation which belongs to Γ_0.

THEOREM 4.5.1. *The series*

$$\sum_{f=0}^{\infty} \sum_{\nu=1}^{N_f} | \varphi_\nu^f(\xi)|^2 r^f \tag{4.5.2}$$

converges uniformly for $\xi \in \mathfrak{C}$ *and* $0 \leqq r \leqq r_0 < 1$. *The sum of the series is equal to* $[V(\mathfrak{C})]^{-1}(1-r)^{-n}$, *where* $V(\mathfrak{C})$ *is the volume of* $\mathfrak{C}$.

PROOF. Without loss of generality we can assume that Γ_0 consists of unitary transformations $\eta = \xi U$. The expression (4.5.2) can be transformed as follows:

$$\sum_{f=0}^{\infty} \xi_f \bar{\xi}_f' r^f = \sum_{f=0}^{\infty} \xi^{[f]} H_2^{-1} \overline{\xi^{[f]}}' r^f = \sum_{f=0}^{\infty} \xi^{[f]} \left(\int_{\mathfrak{C}} \overline{\zeta^{[f]}}' \zeta^{[f]} \dot{\zeta} \right)^{-1} \overline{\xi^{[f]}}' r^f$$

$$= \sum_{f=0}^{\infty} \eta^{[f]} \left(\int_{\mathfrak{C}} \overline{\zeta^{[f]}}' \zeta^{[f]} \dot{\zeta} \right)^{-1} \overline{\eta^{[f]}}' r^f = \sum_{f=0}^{\infty} \eta_f \bar{\eta}_f' r^f,$$

whence it is evident that it does not depend on ξ on $\mathfrak{C}$. Hence, integrating over $\mathfrak{C}$, we obtain

$$\frac{1}{V(\mathfrak{C})} \cdot \int_{\mathfrak{C}} \sum_{f=0}^{\infty} \eta_f \bar{\eta}_f' r^f \cdot \dot{\eta} = \frac{1}{V(\mathfrak{C})} \sum_{f=0}^{\infty} \binom{n+f-1}{f} r^f = \frac{1}{V(\mathfrak{C})} \cdot (1-r)^{-n}, \tag{4.5.3}$$

and the theorem is proved.

From the preceding theorem we obtain by means of the Cauchy-Schwarz inequality the following proposition.

THEOREM 4.5.2. *For* ξ *and* η *on* $\mathfrak{C}$ *and* $0 \leqq r \leqq r_0 < 1$, *the series*

$$\sum_{f=0}^{\infty} \sum_{\nu=1}^{N_f} \varphi_\nu^f(\xi) \overline{\varphi_\nu^f(\eta)} r^f \tag{4.5.4}$$

converges uniformly.

THEOREM 4.5.3. *Let* $\mathfrak{R}$ *be a star-shaped circular domain. We shall denote by* $\mathfrak{R}(r)$ *the domain obtained from* $\mathfrak{R}$ *by a similarity transformation with similarity coefficient* r, $0 < r < 1$. *For* $z \in \mathfrak{R}(r)$ *and* $\xi \in \mathfrak{C}$, *series* (4.5.1) *converges uniformly.*

PROOF. Since a function which is analytic in $\mathfrak{R}$ attains its maximum modulus on $\mathfrak{C}$, it follows that

$$\left|\sum_{f=m}^{m'}\sum_{\nu=1}^{N_f}\varphi_\nu^f(z)\overline{\varphi_\nu^f(\xi)}\right| = \left|\sum_{f=m}^{m'}\sum_{\nu=1}^{N_f}\varphi_\nu^f\left(\frac{z}{r}\right)\overline{\varphi_\nu^f(\xi)}\, r^f\right|$$

$$\leqslant \max_{\zeta\in\mathfrak{C}}\left|\sum_{f=m}^{m'}\sum_{\nu=1}^{N_f}\varphi_\nu^f(\zeta)\overline{\varphi_\nu^f(\xi)}\, r^f\right|$$

[$\varphi_\nu^f(z)$ are homogeneous forms of z of order f]. For $m, m'\to\infty$ the right-hand side of the above equality tends to zero by Theorem 4.5.2. Thus the theorem is proved.

4.6. **The Cauchy formula.** The following two theorems were essentially already demonstrated in full in the preceding section. However, in view of their importance we shall formulate them separately.

THEOREM 4.6.1. *Suppose $\mathfrak{C}$ is the characteristic manifold of the domain $\mathfrak{R}$, satisfying the conditions of Theorem 4.5.3, and $f(\xi)$ is a continuous function on $\mathfrak{C}$. Then the integral*

$$\varphi(z) = \int_{\mathfrak{C}} H(z, \overline{\xi}) f(\xi)\dot{\xi} \tag{4.6.1}$$

represents an analytic function which is regular in $\mathfrak{R}$. Hence, if $f(z)$ is a function which is analytic in $\mathfrak{R}$ and on the boundary of $\mathfrak{R}$, then

$$f(z) = \int_{\mathfrak{C}} H(z, \overline{\xi}) f(\xi)\dot{\xi}.$$

This theorem follows from the uniform convergence of the series for $H(z, \overline{\xi})$ (Theorem 4.5.3).

THEOREM 4.6.2. *Suppose $\{\varphi_\nu(\xi)\}$ is an orthonormal system of functions on $\mathfrak{C}$, with the following properties:*

(1) *$\varphi_\nu(z)$ are analytic functions which are regular in the domain $\mathfrak{R}$ and on its boundary.*

(2) *The series*

$$H_1(z, \overline{\xi}) = \sum_{\nu=0}^{\infty}\varphi_\nu(z)\overline{\varphi_\nu(\xi)}$$

converges uniformly for $\xi\in\mathfrak{C}$ and z lying in any closed subdomain of $\mathfrak{R}$.

(3) *Any function $f(z)$ which is analytic in the domain $\mathfrak{R}$ and on its boundary can be expanded in a series*

$$f(z)=\sum_{\nu=0}^{\infty}a_\nu\varphi_\nu(z),$$

which converges uniformly in any closed subdomain of $\mathfrak{R}$.
Then

$$H_1(z,\bar{\xi})=H(z,\bar{\xi}).$$

Proof. If the orthonormal system $\{\varphi_\nu(z)\}$ satisfies the above conditions, then it is easy to show that Theorem 4.6.1 holds. Rephrasing, if $f(z)$ is analytic in $\mathfrak{R}$ and on its boundary, then

$$f(z)=\int_{\mathfrak{C}}H_1(z,\bar{\xi})f(\xi)\dot{\xi}. \qquad (4.6.2)$$

From the principle of the maximum modulus it follows that the series

$$H(z,\bar{w})=\sum_{f,i}\varphi_i^f(z)\overline{\varphi_i^f(\omega)}$$

converges uniformly for z lying in any closed subdomain $\mathfrak{R}$, and for w lying in $\mathfrak{R}$ or on its boundary. But $H_1(z,\bar{w})$ has the same properties. Hence

$$H(z,\bar{w})=\int_{\mathfrak{C}}H_1(z,\bar{\xi})H(\xi,\bar{w})\dot{\xi}$$

$$=\overline{\int_{\mathfrak{C}}H_1(\bar{z},\xi)H(\bar{\xi},w)\dot{\xi}}=\overline{H_1(\bar{z},w)}=H_1(z,\bar{w}),$$

from which it is evident that $H_1(z,\bar{w})\equiv H(z,\bar{w})$. For $w=\xi$, we obtain the assertion of our theorem.

Theorem 4.6.3. *Let us assume, in addition to the previous conditions, that the domain* $\mathfrak{R}$ *is transitive, and that* $\mathfrak{C}$ *has real dimensionality* n. *Then*

$$H(z,\bar{\xi})=\frac{1}{V(\mathfrak{C})}B^{\frac{1}{2}}(z,z,U)\overline{B^{\frac{1}{2}}(\xi,z,U)},$$

where $B(z,a,U)$ *is the value of the Jacobian of a transformation of the group* Γ *which carries the point* a *into the origin.*

Proof. Let

$$w=f(z,a,U) \qquad (4.6.3)$$

be a transformation of the group Γ which carries the point a into the origin.

This transformation carries $\mathfrak{C}$ into itself. We shall set

$$\zeta = f(\xi, a, U). \tag{4.6.4}$$

Then

$$\dot{\zeta} = B(\xi, a, U)\dot{\xi}. \tag{4.6.5}$$

(This formula is valid, since the real dimensionality of $\mathfrak{C}$ is equal to the complex dimensionality of $\mathfrak{R}$, thus enabling us to effect the passage from the boundary to the interior of the domain by a simple replacement of real parameters by complex ones.)

We know that on $\mathfrak{C}$ there exists an orthonormal system $\{\varphi_\nu^f(\xi)\}$ which we shall denote, for the sake of simplicity, by $\{\varphi_\nu(\xi)\}$. Then

$$\int_{\mathfrak{C}} \varphi_\mu(\zeta)\overline{\varphi_\nu(\zeta)}\dot{\zeta} = \int_{\mathfrak{C}} \varphi_\mu(f(\xi))\overline{\varphi_\nu(f(\xi))}\,|B(\xi, a, U)|\,\dot{\xi} = \delta_{\mu\nu}.$$

Hence the system

$$\psi_\mu(\xi) = \varphi_\mu(f(\xi, a, U))\,B^{\frac{1}{2}}(\xi, a, U)$$

is also orthonormal. Let us prove that the system $\{\psi_\mu(\xi)\}$ satisfies the conditions of Theorem 4.6.2. It is evident that $\psi_\mu(z)$ is analytic in $\mathfrak{R}$ and on its boundary, and

$$\sum_{\nu=0}^{\infty} \psi_\nu(z)\overline{\psi_\nu(\xi)} = \sum_{\nu=0}^{\infty} \varphi_\nu(w)\overline{\varphi_\nu(\zeta)} \cdot B^{\frac{1}{2}}(z, a, U)\overline{B^{\frac{1}{2}}(\xi, a, U)}. \tag{4.6.6}$$

Let us denote by

$$z = f^{-1}(w, a, U)$$

the inverse transformation of (4.6.3). To any function $\psi(z)$ which is analytic in $\mathfrak{R}$ and on its boundary, we shall set in correspondence the function

$$\varphi(w) = \psi\left(f^{-1}(w, a, U)\right) B^{-\frac{1}{2}}(z, a, U),$$

also analytic in $\mathfrak{R}$ and on its boundary. Since $\varphi(w)$ can be expanded in the series

$$\varphi(w) = \sum_{\nu=0}^{\infty} a_\nu \varphi_\nu(w),$$

we also obtain

$$\psi(z) = \sum_{\nu=0}^{\infty} a_\nu \psi_\nu(z).$$

This shows that condition (3) of Theorem 4.6.2 is satisfied. Thus, from (4.6.6) follows

$$H(z, \bar{\xi}) = H(w, \bar{\zeta}) \cdot B^{\frac{1}{2}}(z, a, U) \overline{B^{\frac{1}{2}}(\xi, a, U)}.$$

Since $H(0, \bar{\zeta}) = [V(\mathfrak{C})]^{-1}$, we obtain the assertion of the theorem after replacing a by z.

REMARK. The orthonormal system $\{\psi_\mu(\xi)\}$ considered above is not complete on $\mathfrak{C}$. There also exists a complete orthonormal system on $\mathfrak{C}$ (see H. Weyl [1]); it can be obtained by supplementing $\{\psi_\mu(\xi)\}$ by some system of functions $\psi_{-\nu}(\xi)$, $\nu = 1, 2, \cdots$. Here we are considering completeness in the space of functions which are continuous on $\mathfrak{C}$, i.e., if $g(\xi)$ is a continuous function on $\mathfrak{C}$, then from

$$\int_{\mathfrak{C}} g(\xi) \overline{\varphi_\nu(\xi)} \dot{\xi} = 0 \qquad (\nu = \pm 1, \pm 2, \ldots)$$

it follows that $g(\xi) \equiv 0$.

4.7. **The Cauchy kernels for classical domains.** By applying Theorem 4.6.3 it is possible to directly obtain the Cauchy kernels for classical domains.

1°. In $\mathfrak{R}_{\mathrm{I}}$, the characteristic manifold $\mathfrak{C}_{\mathrm{I}}$ is determined by the condition $U\bar{U}' = I$. Let us first assume that $m = n$. Then the dimensionality of the characteristic manifold equals half the dimensionality of $\mathfrak{R}_{\mathrm{I}}$, and hence

$$H(Z, \bar{U}) = \frac{1}{V(\mathfrak{C}_{\mathrm{I}})} \cdot \{\det(I - Z\bar{U}')\}^{-n}, \tag{4.7.1}$$

where by Theorem 3.1.1

$$V(\mathfrak{C}_{\mathrm{I}}) = \frac{(2\pi)^{\frac{n(n+1)}{2}}}{1!\,2! \ldots (n-1)!}.$$

If $m \neq n$, then, assuming for definiteness that $m < n$, we have

$$H(Z, \bar{U}) = \frac{1}{V(\mathfrak{C}_{\mathrm{I}})} [\det(I - Z\bar{U}')]^{-n}, \tag{4.7.2}$$

where (see §5.4)

$$V(\mathfrak{C}_{\mathrm{I}}) = \frac{(2\pi)^{mn - \frac{m(m-1)}{2}}}{(n-m)!\,(n-m+1)! \ldots (n-1)!}.$$

It is simpler to obtain expression (4.7.2) from (4.7.1) than directly from Theorem 4.6.3. In fact, we have

$$f(Z)=\frac{1}{V(\mathfrak{U}_n)}\int_{\mathfrak{U}_n} f(U_n)[\det(I-Z\bar{U}'_n)]^{-n}\dot{U}_n, \tag{4.7.3}$$

where the integral is taken over the set of all unitary matrices of order n. Setting

$$Z=\begin{pmatrix} Z_1 \\ 0 \end{pmatrix}, \text{ where } Z_1 \text{ — } m\times n\text{- matrix}, \quad U_n=\begin{pmatrix} U_{mn} \\ V \end{pmatrix},$$

we obtain

$$f\left[\begin{pmatrix} Z_1 \\ 0 \end{pmatrix}\right]=\frac{1}{V(\mathfrak{U}_n)}\int_{U_{mn}}\int_V f\left[\begin{pmatrix} U_{mn} \\ V \end{pmatrix}\right][\det(I-Z_1\bar{U}'_{mn})]^{-n}\dot{U}_{mn}\dot{V}$$

$$=\frac{1}{V(\mathfrak{U}_n)}\int_{U_{mn}}[\det(I-Z_1\bar{U}'_{mn})]^{-n}\dot{U}_{mn}\int_V f\left[\begin{pmatrix} U_{mn} \\ V \end{pmatrix}\right]\dot{V}, \tag{4.7.4}$$

where V runs through the set of all $(n-m)\times n$ matrices which satisfy the conditions

$$V\bar{V}'=I^{(n-m)}, \quad U_{mn}\bar{V}'=0. \tag{4.7.5}$$

For a fixed matrix U_{mn} we also find two unitary matrices, namely, the $m\times n$ matrix P and the $n\times n$ matrix Q, such that

$$PU_{mn}Q=(I^{(m)}, 0).$$

Hence it follows that $(I^{(m)}, 0)\overline{Q'V'}=0$, i.e., $VQ=(0, W)$, where W is the unitary matrix of order $n-m$. Therefore

$$\int_V f\left[\begin{pmatrix} U_{mn} \\ V \end{pmatrix}\right]\dot{V}=\int_W f\left[\begin{pmatrix} U_{mn} \\ (0, W) \end{pmatrix}\right]\dot{W}.$$

By (4.7.3), where n has been replaced by $n-m$, and $Z=0$, we have

$$\int_V f\left[\begin{pmatrix} U_{mn} \\ V \end{pmatrix}\right]\dot{V}=V(\mathfrak{U}_{n-m})f\left[\begin{pmatrix} U_{mn} \\ 0 \end{pmatrix}\right].$$

From (4.7.4) it now follows that

$$f\left[\begin{pmatrix} Z_1 \\ 0 \end{pmatrix}\right]=\frac{V(\mathfrak{U}_{n-m})}{V(\mathfrak{U}_n)}\int_{U_{mn}} f\left[\begin{pmatrix} U_{mn} \\ 0 \end{pmatrix}\right][\det(I-Z\bar{U}'_{mn})]^{-n}\dot{U}_{mn},$$

from which we obtain formula (4.7.2) for the Cauchy kernel.

2°. The characteristic manifold of the domain $\mathfrak{R}_{\text{II}}$ is determined from the condition $U\overline{U}=I$, i.e., it coincides with the set of symmetric unitary matrices. Since the dimensionality of $\mathfrak{C}_{\text{II}}$ is half the dimensionality of $\mathfrak{R}_{\text{II}}$, Theorem 4.6.3 yields at once

$$H(Z,\overline{U})=\frac{1}{V(\mathfrak{C}_{\text{II}})}\cdot[\det(I-Z\overline{U})]^{-\frac{n+1}{2}}, \tag{4.7.6}$$

where (see §3.5)

$$V(\mathfrak{C}_{\text{II}})=2^{\frac{n(3n+1)}{4}}\cdot\pi^{\frac{n(n+1)}{4}}\cdot\frac{\Gamma\left(\frac{1}{2}\right)}{\Gamma\left(\frac{n+1}{2}\right)}\cdot\prod_{\nu=1}^{n-1}\frac{\Gamma\left(\frac{n-\nu}{2}+1\right)}{\Gamma(n-\nu+1)}.$$

3°. The characteristic manifold of the domain $\mathfrak{R}_{\text{III}}$ coincides with the set of matrices K defined by the equality (3.6.16).[14]

For even n

$$H(Z,\overline{K})=\frac{1}{V(\mathfrak{C}_{\text{III}})}\cdot[\det(I+Z\overline{K})]^{-\frac{n-1}{2}}, \tag{4.7.7}$$

where (see §3.7)

$$V(\mathfrak{C}_{\text{III}})=\frac{1}{2^{\nu-1}}\cdot(8\pi)^{\frac{n(n-1)}{4}}\cdot\prod_{s=1}^{n-1}\frac{\Gamma\left(\frac{s}{2}\right)}{\Gamma(s)},\quad \nu=\frac{n}{2}.$$

For odd n

$$H(Z,\overline{K})=\frac{1}{V(\mathfrak{C}_{\text{III}})}\cdot\{\det(I+Z\overline{K})\}^{-\frac{n}{2}}, \tag{4.7.8}$$

where

$$V(\mathfrak{C}_{\text{III}})=2\pi\frac{1}{2^{\nu-1}}\cdot(8\pi)^{\frac{n(n+1)}{4}}\cdot\prod_{s=1}^{n}\frac{\Gamma\left(\frac{s}{2}\right)}{\Gamma(s)},\quad \nu=\frac{n+1}{2}.$$

Equality (4.7.7) is obtained directly from Theorem 4.6.3, whereas equality (4.7.8) will now be derived from (4.7.7). Replacing n by $n+1$, we shall write Cauchy's formula with the kernel (4.7.7)

$$f(Z)=\frac{1}{V}\cdot\int_K f(K)\{\det(I+Z\overline{K})\}^{-\frac{n}{2}}\dot{K}, \tag{4.7.9}$$

[14] Note by the translator into English. This equality, found in Chapter III, defines the set of skew-symmetric matrices K.

where Z and K are matrices of order $(n+1)$, and $V=V(\mathfrak{C}_{\mathrm{III}}^{(n+1)})$. We shall set

$$Z=\begin{pmatrix}0 & 0\\ 0 & Z_1\end{pmatrix}, \quad \text{where } Z_1 \text{ is an } n\times n \text{ matrix,}$$

and accordingly:

$$K=\begin{pmatrix}0 & k\\ -k' & K_1\end{pmatrix},$$

where K_1 is an $n\times n$ matrix and k is a $1\times n$ matrix. Then

$$f\left(\begin{pmatrix}0 & 0\\ 0 & Z_1\end{pmatrix}\right)=\frac{1}{V}\int_{K_1}\{\det(I+Z_1\bar{K}_1)\}^{-\frac{n}{2}}\dot{K}_1\int_k f\left(\begin{pmatrix}0 & k\\ -k' & K_1\end{pmatrix}\right)\dot{k}. \quad (4.7.10)$$

Since K is a unitary matrix, it follows that

$$k\bar{k}'=1, \quad k\bar{K}_1'=0, \quad k'\bar{k}+K_1\bar{K}_1'=I^{(n)}.$$

Since the matrix $I-K_1\bar{K}_1'=k'\bar{k}$ has rank one and, furthermore,

$$(I-K_1\bar{K}_1')^2=k'\bar{k}k'\bar{k}=k'\bar{k}=I-K_1\bar{K}_1',$$

a unitary matrix U can be found such that

$$U(I-K_1\bar{K}_1')\bar{U}'=[1,\ 0,\ \ldots,\ 0],$$

i.e.,

$$UK_1\bar{K}_1'\bar{U}'=[0,\ 1,\ \ldots,\ 1]=\begin{pmatrix}0 & 0\\ 0 & F\end{pmatrix}\overline{\begin{pmatrix}0 & 0\\ 0 & F\end{pmatrix}}',$$

where

$$F=\begin{pmatrix}0 & 1\\ -1 & 0\end{pmatrix}\dotplus\cdots\dotplus\begin{pmatrix}0 & 1\\ -1 & 0\end{pmatrix}.$$

Since the element in the upper left-hand corner of the matrix $UK_1\bar{K}_1'\bar{U}'$ is equal to zero, the first row of the matrix UK_1 consists of zeros only, i.e.,

$$UK_1=\begin{pmatrix}0,\ 0,\ \ldots,\ 0\\ *\end{pmatrix}.$$

From the fact that UK_1U' is a skew-symmetric matrix it follows that

$$UK_1U'=\begin{pmatrix}0 & 0\\ 0 & Q\end{pmatrix},$$

where Q is a skew-symmetric matrix of order $(n-1)$ and $Q\bar{Q}'=I=F\bar{F}'$.

But if Q is a unitary skew-symmetric matrix, then a unitary matrix V_0 can be found such that $V_0 Q V_0' = F$. Hence it is also possible to find a unitary matrix U_0 such that

$$\overline{U}_0' K_1 \overline{U}_0 = \begin{pmatrix} 0 & 0 \\ 0 & F \end{pmatrix}.$$

Setting $h = k\overline{U}_0$, we obtain

$$h \begin{pmatrix} 0 & 0 \\ 0 & F \end{pmatrix}' = k\overline{U}_0 U_0' \overline{K}_1' U_0 = 0.$$

This means that

$$h = [e^{i\theta}, 0, \ldots, 0].$$

Thus the inner integral in formula (4.7.10) is equal to

$$\int_0^{2\pi} f\left[\begin{pmatrix} 0 & hU_0' \\ -U_0 h' & K_1 \end{pmatrix}\right] d\theta = 2\pi f\left[\begin{pmatrix} 0 & 0 \\ 0 & K_1 \end{pmatrix}\right]$$

and we obtain (4.7.8) from (4.7.10).

(We used the formula

$$\frac{1}{2\pi} \int_0^{2\pi} \varphi(z_1 e^{i\theta}, \ldots, z_n e^{i\theta})\, d\theta = \varphi(0, 0, \ldots, 0),$$

which holds for $\varphi(z)$ analytic in a closed circular domain with its center at the origin and z lying inside the domain.)

4°. The characteristic manifold of the domain $\mathfrak{R}_{IV}$ consists of vectors of the form $e^{i\theta}x$, where $0 \leq \theta \leq \pi$, and $x = (x_1, \cdots, x_n)$ is a real vector which satisfies the condition $xx' = 1$.

$$H(z, \theta, x) = \frac{1}{V(\mathfrak{C}_{IV})\,[(x - e^{-i\theta} z)(x - e^{-i\theta} z)']^{n/2}}, \tag{4.7.11}$$

It is easy to calculate the magnitude of the volume $V(\mathfrak{C}_{IV})$:

$$V(\mathfrak{C}_{IV}) = \frac{2\pi^{\frac{n}{2}+1}}{\Gamma\left(\frac{n}{2}\right)}.$$

4.8. **The Poisson kernel for circular domains.** Suppose that $\mathfrak{R}$, just as in §4.5, is a star-shaped circular domain, and $\mathfrak{C}$ its characteristic manifold, transitive with respect to the group Γ_0 of motions of $\mathfrak{R}$ which leave the origin unchanged. Then, by Theorem 4.6.1, there exists a Cauchy kernel for the domain $\mathfrak{R}$, and Cauchy's formula holds for any function $f(z)$ which is analytic in $\mathfrak{R}$ and on its boundary.

Setting, in particular,

$$f(z) = H(z, \overline{w})\, g(z),$$

where $g(z)$ is an arbitrary function which is analytic in $\mathfrak{R}$ and on its boundary, we have

$$H(z, \overline{w})\, g(z) = \int_{\mathfrak{C}} H(z, \bar{\xi})\, H(\xi, \overline{w})\, g(\xi)\, \dot{\xi}.$$

For $w = z$, we obtain Poisson's formula

$$g(z) = \int_{\mathfrak{C}} P(z, \xi)\, g(\xi)\, \dot{\xi}, \tag{4.8.1}$$

where the kernel

$$P(z, \xi) = \frac{H(z, \bar{\xi})\, H(\xi, \bar{z})}{H(z, \bar{z})} \tag{4.8.2}$$

has occured above under the name of the Poisson kernel of the domain $\mathfrak{R}$.

Up to now we have established that formula (4.8.1) is valid for analytic $g(z)$; yet it can be extended to other classes of functions too (see §5.8). For any continuous function $u(\xi)$ the integral

$$u(z) = \int_{\mathfrak{C}} P(z, \xi)\, u(\xi)\, \dot{\xi} \tag{4.8.3}$$

defines a certain function. It can be proved (see §5.8) that $u(z) \to u(\xi)$ for $z \to \xi$. Functions of the form (4.8.3) we shall call harmonic functions in $\mathfrak{R}$. It is reasonable to expect that if there exists a complete orthonormal system $\{\psi_\nu(\xi)\}$ on $\mathfrak{C}$, then the set of functions which are harmonic in $\mathfrak{R}$ is the closure of the linear span of the system $\{\psi_\nu(z)\}$ (see §5.10).

If $\mathfrak{R}$ satisfies the conditions of Theorem 4.6.3, then the Poisson kernel can be written in the following simple form:

$$P(z, \xi) = \frac{1}{V(\mathfrak{C})} \cdot |B(\xi, z, U)|. \tag{4.8.4}$$

In conclusion let us list the Poisson kernels for the classical domains.

(1) For $\mathfrak{R}_{\mathrm{I}}$

$$P(Z, U) = \frac{1}{V(\mathfrak{C}_{\mathrm{I}})} \cdot \frac{[\det(I - Z\bar{Z}')]^n}{|\det(I - Z\bar{U}')|^{2n}}, \tag{4.8.5}$$

where $U \in \mathfrak{C}_{\mathrm{I}}$. In particular, for $m = n$, one can also write

$$P(Z, U) = \frac{1}{V(\mathfrak{C}_{\mathrm{I}})} \cdot \frac{[\det(I - Z\bar{Z}')]^n}{|\det(Z - U)|^{2n}}.$$

(2) For $\mathfrak{R}_{\mathrm{II}}$

$$P(Z, U) = \frac{1}{V(\mathfrak{C}_{\mathrm{II}})} \cdot \frac{[\det(I - Z\bar{Z})]^{\frac{n+1}{2}}}{|\det(I - Z\bar{U})|^{n+1}}, \tag{4.8.6}$$

where $U \in \mathfrak{C}_{\mathrm{II}}$.

(3) For $\mathfrak{R}_{\mathrm{III}}$ with even n

$$P(Z, K) = \frac{1}{V(\mathfrak{C}_{\mathrm{III}})} \cdot \frac{[\det(I + Z\bar{Z})]^{\frac{n-1}{2}}}{|\det(I + Z\bar{K})|^{n-1}}, \tag{4.8.7}$$

and with odd n

$$P(Z, K) = \frac{1}{V(\mathfrak{C}_{\mathrm{III}})} \cdot \frac{[\det(I + Z\bar{Z})]^{\frac{n}{2}}}{|\det(I + Z\bar{K})|^{n}}. \tag{4.8.8}$$

In both cases $K \in \mathfrak{C}_{\mathrm{III}}$.

(4) For $\mathfrak{R}_{\mathrm{IV}}$

$$P(z, \xi) = \frac{1}{V(\mathfrak{C}_{\mathrm{IV}})} \cdot \frac{(1 + |zz'|^2 - 2\bar{z}z')^{\frac{n}{2}}}{|(z - \xi)(z - \xi)'|^n}, \tag{4.8.9}$$

where $\xi \in \mathfrak{C}_{\mathrm{IV}}$.

CHAPTER V

HARMONIC ANALYSIS IN THE SPACE OF RECTANGULAR MATRICES

5.1. **Orthogonal systems in the space of rectangular matrices.** Let us denote by $x=(x_1, x_2, \cdots, x_m)$ a vector in m-dimensional complex space. By $x^{[f]}$ we shall denote a vector with the components

$$\sqrt{\frac{f!}{s_1! \ldots s_m!}}\, x_1^{s_1} x_2^{s_2} \ldots x_m^{s_m} \qquad (s_1+\ldots+s_m=f), \tag{5.1.1}$$

whose dimensionality is

$$\frac{1}{f!} m(m+1) \ldots (m+f-1). \tag{5.1.2}$$

When the vector x is mapped into the vector y by a linear transformation with matrix P, the vector $x^{[f]}$ is mapped into the vector $y^{[f]}$ by a linear transformation with matrix $P^{[f]}$. The vector $x^{[f]}$ we shall call the fth symmetrized Kronecker power of the vector x, and the matrix $P^{[f]}$ the fth symmetrized Kronecker power of the matrix P.

It is evident that expression (5.1.1) contains all the monomials of degree f, i.e., any homogeneous form of $x_1, x_2, \cdots, x_m$ of order f is a linear combination of expressions of the form (5.1.1). Any polynomial of $x_1, x_2, \cdots, x_m$ is a linear combination of expressions of the form (5.1.1), where f assumes the values $0, 1, 2, \cdots$.

The relation

$$\operatorname{Sp}\left((P\times Q)^{[f]}\right) = \sum_{\substack{f_1+\ldots+f_m=f \\ f_1\geqslant f_2\geqslant\ldots\geqslant f_m\geqslant 0}} \chi_{f_1,\ldots,f_m}(P)\cdot\chi_{f_1,\ldots,f_m,0,\ldots,0}(Q), \quad \begin{array}{l} P\in GL(m), \\ Q\in GL(n), \end{array} \tag{5.1.3}$$

which constitutes the assertion of Theorem 1.4.1, can be formulated in the language of the theory of representations as follows: the fth symmetrized Kronecker power of the Kronecker product of $GL(m)$ by $GL(n)$ can be decomposed into a direct sum of irreducible components, where each of the components is met precisely once. These irreducible components are the Kronecker product of the representation of $GL(m)$ with signature $(f_1, f_2, \cdots, f_m)$ and of the representation of $GL(n)$ with signature $(f_1, \cdots, f_m, 0, \cdots, 0)$. (We assume everywhere that $n \geqq m$.)

For elucidation let us consider the transformation

$$W = P'ZQ, \tag{5.1.4}$$

where the variables undergoing the transformation are the $m \times n$ matrices Z and W, whereas the $m \times m$ matrix P and the $n \times n$ matrix Q determine the transformation. Let us arrange the elements of the matrices Z and W in the form of vectors

$$\begin{aligned} z &= (z_{11}, \ldots, z_{1n}, z_{21}, \ldots, z_{2n}, \ldots, z_{m1}, \ldots, z_{mn}), \\ w &= (w_{11}, \ldots, w_{1n}, w_{21}, \ldots, w_{2n}, \ldots, w_{m1}, \ldots, w_{mn}). \end{aligned} \tag{5.1.5}$$

The transformation (5.1.4) of the matrix Z into the matrix W induces a transformation of the vector z into the vector w. The matrix of this transformation is equal to the Kronecker product of the matrices P and Q, i.e., $P \times Q$, whereas the matrix of the corresponding transformation of $z^{[f]}$ into $w^{[f]}$ is $(P \times Q)^{[f]}$. The above-mentioned theorem states that the subspace $z^{[f]}$, invariant under this transformation, is decomposed into a direct sum of subspaces with dimensions

$$q(f_1, \ldots, f_m) = N(f_1, \ldots, f_m) \cdot N(f_1, \ldots, f_m, 0, \ldots, 0). \tag{5.1.6}$$

Let us denote by

$$\psi^{(i)}_{f_1, \ldots, f_m}(z), \quad i = 1, 2, \ldots, q(f_1, \ldots, f_m), \tag{5.1.7}$$

the components of $z^{[f]}$. When Z is mapped into W by the transformation (5.1.4), $\psi^{(i)}_{f_1, \cdots, f_m}(z)$ is mapped into a linear combination of $\psi^{(i)}_{f_1, \cdots, f_m}(w)$ by means of the matrix

$$A_{f_1, \ldots, f_m}(P) \times A_{f_1, \ldots, f_m, 0, \ldots, 0}(Q). \tag{5.1.8}$$

The group of motions $\mathfrak{R}_{\mathrm{I}}$ which leave the origin unchanged consists of transformations of the form

$$W = U'ZV,$$

where U and V are unitary matrices of order m and n, respectively. For different $(f_1, \cdots, f_m)$ the representations

$$A_{f_1,\ldots,f_m}(U)\times A_{f_1,\ldots,f_m,0,\ldots,0}(V)$$

are not equivalent. Hence, by Schur's lemma,

$$\int\limits_{I-Z\bar{Z}'>0}\psi_f^{(i)}(z)\,\overline{\psi_g^{(j)}(Z)}\cdot\dot{Z}=\delta_{fg}\cdot\delta_{ij}\cdot\rho_f \tag{5.1.9}$$

[f stands for $(f_1,f_2,\cdots,f_m)$], where ρ_f does not depend on i. Thus the set of functions

$$\{\psi_f^{(i)}(Z)\}_{i,f}$$

forms an orthogonal system. From the results of Chapter IV it follows that this system is complete. We have still to compute the integrals

$$\rho_f=\int\limits_{I-Z\bar{Z}'>0}|\psi_f^{(i)}(Z)|^2\,\dot{Z}. \tag{5.1.10}$$

For this purpose we shall first refine the process of obtaining the functions $\psi_f^{(i)}(Z)$. The vector obtained from the matrix Z is transformed by the matrix (5.1.8) when the matrix Z undergoes the transformation (5.1.4). For the diagonal matrix $\Lambda=[\lambda_1,\cdots,\lambda_n]$ the matrix $A_{f_1,\cdots,f_n}(\Lambda)$ is also diagonal. By arranging the rows and columns in suitable order, we can assume that

$$\lim_{\substack{\lambda_{m+1}\to 0\\ \cdots\cdots\\ \lambda_n\to 0}} A_{f_1,\ldots,f_m,0,\ldots,0}(\Lambda)=\begin{pmatrix}A_{f_1,\ldots,f_m}[\lambda_1,\ldots,\lambda_m] & 0\\ 0 & 0\end{pmatrix}.$$

This means that also for

$$X=\begin{pmatrix}X_1^{(m)} & 0\\ 0 & 0\end{pmatrix}$$

we have

$$A_{f_1,\ldots,f_m,0,\ldots,0}(X)=\begin{pmatrix}A_{f_1,\ldots,f_m}(X_1^{(m)}) & 0\\ 0 & 0\end{pmatrix}. \tag{5.1.11}$$

Let us now complete the $m\times n$ matrix Z, $n\geq m$, to a square matrix of order n

$$\begin{pmatrix}Z\\ 0\end{pmatrix},$$

hence

$$A_{f_1,\dots,f_m,0,\dots,0}\left[\begin{pmatrix}P' & 0\\ 0 & 0\end{pmatrix}\right]A_{f_1,\dots,f_m,0,\dots,0}\left[\begin{pmatrix}Z\\ 0\end{pmatrix}\right]$$
$$\times A_{f_1,\dots,f_m,0,\dots,0}(Q)=A_{f_1,\dots,f_m,0,\dots,0}\left[\begin{pmatrix}P'ZQ\\ 0\end{pmatrix}\right].$$

From this it follows that

$$A_{f_1,\dots,f_m,0,\dots,0}\left[\begin{pmatrix}Z\\ 0\end{pmatrix}\right]=\begin{pmatrix}L(Z)\\ 0\end{pmatrix} \tag{5.1.12}$$

and

$$A_{f_1,\dots,f_m,0,\dots,0}(P')\cdot L(Z)\cdot A_{f_1,\dots,f_m,0,\dots,0}(Q)=L(P'ZQ). \tag{5.1.13}$$

The matrix $L(Z)$ has $N(f_1,\cdots,f_m)$ rows and $N(f_1,\cdots,f_m,0,\cdots,0)$ columns, its elements being forms of order f of elements of Z. If the elements of the matrix $L(Z)$ are arranged in the form of the vector $l(Z)$, then under a corresponding transformation of Z into W, the vector $l(Z)$ is transformed into the vector $l(W)$ by the matrix

$$A_{f_1,\dots,f_m}(P)\times A_{f_1,\dots,f_m,0,\dots,0}(Q).$$

Hence we can assume that the $\psi^{(i)}_{f_1,\cdots,f_m}(Z)$ coincide with the elements of the matrix $L(Z)$, arranged in a certain order. Therefore

$$\sum_{i=1}^{q(f_1,\dots,f_m)}\left|\psi^{(i)}_{f_1,\dots,f_m}(Z)\right|^2=\mathrm{Sp}\,\{L(Z)\overline{L(Z)}'\},$$

and we have

$$q(f_1,\dots,f_m)\rho_{f_1,\dots,f_m}=\int_{I-Z\bar Z'>0}\mathrm{Sp}\,\{L(Z)\overline{L(Z)}'\}\,\dot Z$$
$$=\int_{I-Z\bar Z'>0}\mathrm{Sp}\left[A_{f_1,\dots,f_m,0,\dots,0}\begin{pmatrix}Z\\ 0\end{pmatrix}\overline{A_{f_1,\dots,f_m,0,\dots,0}\begin{pmatrix}Z\\ 0\end{pmatrix}'}\right]\dot Z$$
$$=\int_{I-Z\bar Z'>0}\mathrm{Sp}\,[A_{f_1,\dots,f_m}(Z\bar Z')]\,\dot Z=\int_{I-Z\bar Z'>0}\chi_{f_1,\dots,f_m}(Z\bar Z')\,\dot Z. \tag{5.1.14}$$

5.2. **Integrals of functions which are invariant under the transformations** $Z\to\Gamma Z\Gamma^{-1}$. Let us consider the integral of the form,

$$\Omega=\int_{I-Z\bar Z'>0}\chi(Z\bar Z')\,\dot Z, \tag{5.2.1}$$

where $\chi(W)$ is a function which satisfies the condition

$$\chi(\Gamma W \Gamma^{-1}) = \chi(W) \tag{5.2.2}$$

for any nonsingular square matrix Γ.

Suppose at first that $m=n$. Then by Theorem 3.4.4 we have

$$\Omega = \frac{2^{-n^2}}{n} \omega_n \omega'_n \int_0^1 \cdots \int_0^1 \chi([\lambda_1, \ldots, \lambda_n]) D^2(\lambda_1, \ldots, \lambda_n)\, d\lambda_1 \ldots d\lambda_n, \tag{5.2.3}$$

where

$$\omega_n = V(\mathfrak{U}_n), \quad \omega'_n = V([\mathfrak{U}_n]).$$

For the case $\chi = \chi_f$, the integral (5.2.3) can be computed by means of the following theorem.

THEOREM 5.2.1.

$$\int_0^1 \cdots \int_0^1 \det\left|\lambda_j^{l_i}\right|_{i,j=1}^n \cdot \det\left|\lambda_j^{m_i}\right|_{i,j=1}^n (\lambda_1 \ldots \lambda_n)^{a-1}\, d\lambda_1 \ldots d\lambda_n$$
$$= n!\, \frac{D(l_1, \ldots, l_n) D(m_1, \ldots, m_n)}{\prod_{j=1}^n \prod_{i=1}^n (a + l_j + m_i)}. \tag{5.2.4}$$

PROOF. As the integrand function is

$$(\lambda_1 \ldots \lambda_n)^{a-1} \sum_j \delta^{1 \ldots n}_{j_1 \ldots j_n} \lambda_{j_1}^{l_1} \ldots \lambda_{j_n}^{l_n} \cdot \sum_s \delta^{1 \ldots n}_{s_1 \ldots s_n} \lambda_{s_1}^{m_1} \ldots \lambda_{s_n}^{m_n}$$
$$= (\lambda_1 \ldots \lambda_n)^{a-1} \sum_j \lambda_{j_1}^{l_1} \ldots \lambda_{j_n}^{l_n} \cdot \sum_k \delta^{1 \ldots n}_{k_1 \ldots k_n} \lambda_{j_1}^{m_{k_1}} \ldots \lambda_{j_n}^{m_{k_n}},$$

it follows that, by integrating with respect to λ_j from zero to unity, we shall find that the integral in the left-hand side of (5.2.4) is

$$\sum_j \sum_s \delta^{1 \ldots n}_{s_1 \ldots s_n} \int_0^1 \cdots \int_0^1 \lambda_{j_1}^{l_1 + m_{s_1} + a - 1} \ldots \lambda_{j_n}^{l_n + m_{s_n} + a - 1}\, d\lambda_1 \ldots \delta\lambda_n$$
$$= \sum_j \sum_s \delta^{1 \ldots n}_{s_1 \ldots s_n} \frac{1}{l_1 + m_{s_1} + a} \cdots \frac{1}{l_n + m_{s_n} + a}$$
$$= \sum_j \det\left(\frac{1}{l_i + m_k + a}\right)_1^n = n! \det\left(\frac{1}{l_i + m_k + a}\right)_1^n;$$

the proof is completed by using Theorem 1.1.3.

In particular, setting

$$m_1=n-1,\ m_2=n-2,\ \ldots,\ m_n=0,\ a=1,$$

we obtain

$$\int_0^1\cdots\int_0^1 \chi_f([\lambda_1,\ \ldots,\ \lambda_n])\, D^2(\lambda_1,\ \ldots,\ \lambda_n)\, d\lambda_1\ \ldots\ d\lambda_n$$

$$=n!((n-1)!\ \ldots\ 2!\,1!)^2\prod_{j=1}^{n}\frac{l_j!}{(l_j+n)!}N(f_1,\ \ldots,\ f_n), \qquad (5.2.5)$$

where $l_j=f_j+n-j$.

Thus, for $m=n$ we have

$$\Omega=\frac{1}{n!}2^{-n^2}\omega_n\omega_n'n!((n-1)!\ \ldots\ 2!\,1!)^2\prod_{j=1}^{n}\frac{l_j!}{(l_j+n)!}N(f_1,\ \ldots,\ f_n)$$

$$=\pi^{n^2}\prod_{j=1}^{n}\frac{l_j!}{(l_j+n)!}N(f_1,\ \ldots,\ f_n). \qquad (5.2.6)$$

Let us now consider the case $m\leqq n$. Suppose

$$X=X^{(n)}=\begin{pmatrix}Z\\ *\end{pmatrix};$$

by Theorem 2.2.2 we have

$$\int_{I-Z\bar Z'>0}\chi(Z\bar Z')[\det(I-Z\bar Z')]^{\lambda}\dot Z$$

$$=c_1\int_{I-X\bar X'>0}\chi(Z\bar Z')[\det(I-Z\bar Z')]^{\lambda-(n-m)}\dot X, \qquad (5.2.7)$$

where

$$c_1=\frac{n!\,(n+1)!\ \ldots\ (2n-m-1)!}{1!\,2!\ \ldots\ (n-m-1)!}\pi^{-n(n-m)},$$

λ being a complex parameter of which it is only required that the integral in the right-hand side of (5.2.7) should converge. Let us express X in the form TU, where U is a unitary matrix and T a triangular (with zeros above the diagonal) matrix with real diagonal elements. By writing T in the form

$$T=\begin{pmatrix}T_1^{(m)} & 0\\ T_2 & T_3\end{pmatrix},$$

we at once obtain $Z=(T_1,0)U$, $Z\bar{Z}'=T_1\bar{T}_1'$, whence (by Theorem 3.4.3)

$$\int_{I-Z\bar{Z}'>0} \chi(Z\bar{Z}')[\det(I-Z\bar{Z}')]^{\lambda}\cdot\dot{Z}$$

$$=\frac{c_1\omega_n}{2^{\frac{n(n-1)}{2}}}\int_{I-T\bar{T}'>0}\chi(T_1\bar{T}_1')\left[\det\left(I-T_1\bar{T}_1'\right)\right]^{\lambda-(n-m)} t_1^{2(n-1)+1}\dots t_{n-1}^3 t_n\cdot\dot{T}. \tag{5.2.8}$$

Let us note that

$$I-T\bar{T}'=\begin{pmatrix} I-T_1\bar{T}_1' & -T_1\bar{T}_2' \\ -T_2\bar{T}_1' & I-T_2\bar{T}_2'-T_3\bar{T}_3' \end{pmatrix}$$

and

$$\begin{pmatrix} I & 0 \\ P & I \end{pmatrix}\begin{pmatrix} I-T_1\bar{T}_1' & -T_1\bar{T}_2' \\ -T_2\bar{T}_1' & I-T_2\bar{T}_2'-T_3\bar{T}_3' \end{pmatrix}\begin{pmatrix} I & \bar{P}' \\ 0 & I \end{pmatrix}$$

$$=\begin{pmatrix} I-T_1\bar{T}_1' & 0 \\ 0 & I-T_2\left(I-\bar{T}_1'T_1\right)^{-1}\bar{T}_2'-T_3\bar{T}_3' \end{pmatrix},$$

where $P=T_2\bar{T}_1'(I-T_1\bar{T}_1')^{-1}$. Since $I-\bar{T}_1'T_1$ is a positive-definite matrix, there exists a matrix Γ such that $(I-\bar{T}_1'T_1)^{-1}=\Gamma\bar{\Gamma}$. Setting $Q=T_2\Gamma$, we obtain

$$\dot{Q}=(\det\Gamma\bar{\Gamma}')^{n-m}\,\dot{T}_2=\{\det(I-T_1\bar{T}_1')\}^{-n+m}\cdot\dot{T}_2.$$

Therefore

$$\int_{I-Z\bar{Z}'>0}\chi(Z\bar{Z}')[\det(I-Z\bar{Z}')]^{\lambda}\,\dot{Z}$$

$$=\frac{c_1\omega_n}{2^{\frac{n(n-1)}{2}}}\int_{1-T_1\bar{T}_1'>0}\chi(T_1\bar{T}_1')\left[\det\left(I-T_1\bar{T}_1'\right)\right]^{\lambda} t_1^{2(n-1)+1}\dots t_m^{2(n-m)+1}\dot{T}_1$$

$$\times\int_{I-Q\bar{Q}'-T_3\bar{T}_3'>0} t_{m+1}^{2(n-m-1)+1}\dots t_n\cdot\dot{Q}\cdot\dot{T}_3. \tag{5.2.9}$$

This holds for any λ. In particular, setting $\lambda=0$, we have

$$\Omega = \frac{c_1 \omega_n}{2^{\frac{n(n-1)}{2}}} \int_{I - T_1 \overline{T}_1' > 0} \chi(T_1 \overline{T}_1') \cdot t_1^{2(n-1)+1} \ldots t_m^{2(n-m)+1} \dot{T}$$

$$\times \int_{I - Q\overline{Q}' - T_3 \overline{T}_3' > 0} t_{m+1}^{2(n-m-1)+1} \ldots t_n \cdot Q \dot{T}_3. \tag{5.2.10}$$

Let us first compute the latter integral. Since $I - T_3 \overline{T}_3'$ is a positive-definite hermitian matrix, there exists a matrix Γ such that $I - T_3 \overline{T}_3' = \Gamma \overline{\Gamma}'$. Let us set $Q = \Gamma R$. Then

$$\dot{Q} = (\det \Gamma \overline{\Gamma}')^{n-m} \cdot \dot{R} = \{\det (I - T_3 \overline{T}_3')\}^{n-m} \cdot \dot{R}.$$

Hence

$$\int_{I - Q\overline{Q}' - T_3 \overline{T}_3' > 0} t_{m+1}^{2(n-m+1)-1} \ldots t_n \cdot \dot{Q} \dot{T}_3$$

$$= \int_{I - R\overline{R}' > 0} \dot{R} \cdot \int_{I - T_3 \overline{T}_3' > 0} t_{m+1}^{2(n-m-1)+1} \ldots t_n [\det (I - T_3 \overline{T}_3')]^{n-m} \cdot \dot{T}_3$$

$$= \frac{V_{m, n-m}}{\omega_{n-m} \cdot 2^{-\frac{(n-m)(n-m-1)}{2}}} \int_{I - Z_3 \overline{Z}_3' > 0} [\det (I - Z_3 \overline{Z}_3')]^{n-m} \cdot \dot{Z}_3 \tag{5.2.11}$$

(by Theorem (3.4.3).

Further,

$$\int_{I - T_1 \overline{T}_1' > 0} \chi(T_1 \overline{T}_1') \cdot t_1^{2(n-1)+1} \ldots t_m^{2(n-m)+1} \cdot \dot{T}_1$$

$$= \frac{2^{\frac{m(m-1)}{2}}}{\omega_m} \int_{I - Z_1 \overline{Z}_1' > 0} \chi(Z_1 \overline{Z}_1') \{\det Z_1 \overline{Z}_1'\}^{n-m} \cdot \dot{Z}_1. \tag{5.2.12}$$

By (5.2.10) we have

$$\Omega = C \int_{I - Z_1 \overline{Z}_1' > 0} \chi(Z_1 \overline{Z}_1') (\det Z_1 \overline{Z}_1')^{n-m} \cdot \dot{Z}_1, \tag{5.2.13}$$

where C is a constant depending only on m and n, whose value we shall calculate as follows.

By setting $\chi(Z_1\overline{Z}_1')=1$, we shall obtain [see (5.2.1)]

$$V_{m,n}=C\int_{I-Z_1\overline{Z}_1'>0}\left(\det Z_1\overline{Z}_1'\right)^{n-m}\cdot\dot{Z}_1$$

$$=C\cdot\pi^{m^2}\frac{(n-m)!\,(n-m+1)!\,\ldots\,(n-1)!}{n!\,(n+1)!\,\ldots\,(n+m-1)!}.$$

But $V_{m,n}=V(\Re_{\rm I})$, so that (by Theorem 2.2.1)

$$C=V_{m,n}\cdot\pi^{-m^2}\frac{n!\,(n+1)!\,\ldots\,(n+m-1)!}{(n-m)!\,(n-m+1)!\,\ldots\,(n-1)!}$$

$$=\pi^{m(n-m)}\frac{1!\,2!\,\ldots\,(m-1)!\;1!\,2!\,\ldots\,(n-1)!}{1!\,2!\,\ldots\,(n+m-1)!}$$

$$\times\frac{n!\,(n+1)!\,\ldots\,(n+m-1)!}{(n-m)!\,(n-m+1)!\,\ldots\,(n-1)!}$$

$$=\pi^{m(n-m)}\frac{1!\,2!\,\ldots\,(m-1)!}{(n-m)!\,(n-m+1)!\,\ldots\,(n-1)!}=V_{m,n-m}.$$

Hence

$$\Omega=V_{m,n-m}\int_{I-Z_1\overline{Z}_1'>0}\chi(Z_1\overline{Z}_1')\left(\det Z_1\overline{Z}_1'\right)^{n-m}\dot{Z}_1. \tag{5.2.14}$$

For $\chi(X)=\chi_{f_1,\cdots,f_m}(X)$, and since

$$\chi_{f_1,\ldots,f_m}(X)\,(\det X)^{n-m}=\chi_{f_1+n-m,\ldots,f_m+n-m}(X),$$

we obtain by (5.2.6) that

$$\int_{I-Z_1\overline{Z}_1'>0}\chi_f(Z_1\overline{Z}_1')\left(\det Z_1\overline{Z}_1'\right)^{n-m}\cdot\dot{Z}_1$$

$$=\frac{\omega_m\omega_m'}{2^{\frac{m(m+1)}{2}}}[1!2!\,\ldots\,(m-1)!]^2\cdot\prod_{j=1}^{m}\frac{(l_j+n-m)!}{(l_j+n)!}N(f_1,\ldots,f_m)$$

$$=\pi^{m^2}\prod_{j=1}^{m}\frac{(l_j+n-m)!}{(l_j+n)!}\cdot N(f_1,\ldots,f_m). \tag{5.2.15}$$

Hence, finally,

$$\Omega=\pi^{m^2}V_{m,n-m}\prod_{j=1}^{m}\frac{(l_j+n-m)!}{(l_j+n)!}N(f_1,\ldots,f_m), \tag{5.2.16}$$

where $l_j = f_j + m - j$.

5.3. **The orthogonal system and the Bergman kernel.** We now have for $\mathfrak{R}_{\mathrm{I}}$ the orthogonal system of functions

$$(\rho_{f_1, \ldots, f_m})^{-\frac{1}{2}} \psi^{(i)}_{f_1, \ldots, f_m}(Z), \qquad i = 1, 2, \ldots, q(f_1, \ldots, f_m),$$

which were obtained by the method of §5.1. This system is complete. Making use of it let us once again obtain the Bergman kernel for $\mathfrak{R}_{\mathrm{I}}$.

In Theorem 1.2.5 we set $\rho = m+1$. Then for $|\lambda_i| < 1$ we have

$$\left(\prod_{i=1}^{m} (1-\lambda_i)\right)^{-n-m} = C_{m+1} \sum_{l_1 > \cdots > l_m \geqslant 0} a_{l_1+n-m} \cdots a_{l_m+n-m} \times N(f_1, \ldots, f_m, 0, \ldots, 0) \cdot \chi_{f_1, \ldots, f_m}([\lambda_1, \ldots, \lambda_m]), \tag{5.3.1}$$

where

$$l_j = f_j + m - j, \quad a_l = \frac{(m+l)!}{m!\, l!}, \quad C_{m+1} = \frac{1}{a_{n-m} \cdots a_{n-1}}.$$

Since $\chi_{f_1, \ldots, f_m}(X)$ and $\det(I - X)$ are invariant with respect to the transformations $X_1 = \Gamma X \Gamma^{-1}$, we have

$$[\det(I - W\bar{Z}')]^{-(n+m)} = C_{m+1} \sum_{l_1 > \cdots > l_m \geqslant 0} a_{l_1+n-m} \cdots a_{l_m+n-m} \times N(f_1, \ldots, f_m, 0, \ldots, 0) \chi_{f_1, \ldots, f_m}(W\bar{Z}') \tag{5.3.2}$$

under the condition that the moduli of the eigenvalues of the matrix $W\bar{Z}'$ should not exceed unity. If W and Z belong to $\mathfrak{R}_{\mathrm{I}}$, i.e.,

$$I - W\bar{W}' > 0, \quad I - Z\bar{Z}' > 0,$$

then it is easy to see that this condition is always satisfied.

We know that

$$K(Z, \bar{W}) = \sum_{f, i} \frac{\psi^{(i)}_{f_1, \ldots, f_m}(Z) \overline{\psi^{(i)}_{f_1, \ldots, f_m}(W)}}{\rho_{f_1, \ldots, f_m}} = \sum_{f} \frac{\chi_{f_1, \ldots, f_m}(Z\bar{W}')}{\rho_{f_1, \ldots, f_m}}. \tag{5.3.3}$$

From (5.1.14) and (5.2.16) it follows that

$$q(f_1, \dots, f_m) \cdot \rho_{f_1, \dots, f_m} = \pi^{m^2} V_{m, n-m} \cdot \prod_{j=1}^{m} \frac{(l_j + n - m)!}{(l_j + n)!} N(f_1, \dots, f_m). \tag{5.3.4}$$

Substituting in (5.3.3), we obtain

$$K(Z, \overline{W}) = \frac{\pi^{-m^2}}{V_{m, n-m}} \sum_{f} \prod_{j=1}^{m} \frac{(l_j + n)!}{(l_j + n - m)!}$$

$$\times N(f_1 \dots, f_m, 0, \dots, 0) \chi_{f_1, \dots, f_m}(Z\overline{W}'),$$

whence we find, by means of (5.3.2),

$$K(Z, \overline{W}) = \frac{\pi^{-m^2}}{V_{m, n-m}} \cdot a_{n-m} \dots a_{n-1} \cdot (m!)^m \{\det(I - Z\overline{W}')\}^{-m-n}$$

$$= c\{\det(I - Z\overline{W}')\}^{-m-n},$$

where

$$c^{-1} = \pi^{m^2} V_{m, n-m} \prod_{i=1}^{m} \frac{(n-i)!}{(m+n-i)!}$$

$$= \pi^{mn} \cdot \frac{1!2! \dots (m-1)!}{n!(n+1)! \dots (n+m-1)!} = V(\mathfrak{R}_{\mathrm{I}}).$$

Thus, we have once more obtained the formula

$$K(Z, \overline{W}) = \frac{1}{V(\mathfrak{R}_{\mathrm{I}})} \{\det(I - Z\overline{W}')\}^{-m-n}.$$

5.4. **Harmonic analysis on the characteristic manifold.** Suppose, for definiteness, that $n \geqq m$, and let U be an $m \times n$ matrix which satisfies the condition

$$U\overline{U}' = I^{(m)}. \tag{5.4.1}$$

The totality of all such matrices forms the manifold $\mathfrak{U}_{m,n}$ which coincides with the set of cosets of the unitary group $\mathfrak{U}_n$ with respect to its subgroup $\mathfrak{U}_{n-m}$, consisting of matrices of the form

$$\begin{pmatrix} I^{(m)} & 0 \\ 0 & U^{(n-m)} \end{pmatrix}.$$

In more detail, let us put

$$U_n = \begin{pmatrix} U_{m,n} \\ P \end{pmatrix}, \quad V_n = \begin{pmatrix} U_{m,n} \\ Q \end{pmatrix}.$$

Then

$$U_n V_n^{-1} = \begin{pmatrix} U_{m,n} \\ P \end{pmatrix} \begin{pmatrix} U_{m,n} \\ Q \end{pmatrix}^{-1} = \begin{pmatrix} I & 0 \\ 0 & U_{n-m} \end{pmatrix}.$$

Hence it follows that if $f(U)$ is defined on $\mathfrak{U}_{m,n}$, i.e., it is constant on the cosets of the group $\mathfrak{U}_n$ with respect to the indicated subgroup, then

$$\int_{\mathfrak{U}_n} f(U)\dot{U}_n = \int_{\mathfrak{U}_{m,n}} f(U)\dot{U} \cdot \int_{\mathfrak{U}_{n-m}} \dot{U}_{n-m}.$$

In this case we define the integral over $\mathfrak{U}_{m,n}$ by the expression

$$\int_{\mathfrak{U}_{m,n}} f(U)\dot{U} = \frac{1}{\omega_{n-m}} \int_{\mathfrak{U}_n} f(U)\dot{U}_n. \tag{5.4.2}$$

As can be readily seen, the volume of $\mathfrak{U}_{m,n}$ is

$$\frac{\omega_n}{\omega_{n-m}} = \frac{(2\pi)^{mn - \frac{m(m-1)}{2}}}{(n-m)! \ldots (n-1)!}.$$

Let us denote (f stands again for $f_1, f_2, \cdots, f_m$):

$$P_f(U) = (\psi_f^{(1)}(U), \ldots, \psi_f^{(q(f))}(U)),$$
$$q(f) = N(f_1, \ldots, f_m) N(f_1, \ldots, f_m, 0, \ldots, 0),$$

where $\psi_f^{(i)}(Z)$ are the above-constructed functions of the orthogonal system in $\mathfrak{R}_{\mathrm{I}}$. Under the transformation $U \to V'^{(m)} U W^{(n)}$ we have

$$P_f(V'^{(m)} U W^{(n)}) = P_f(U) \cdot (A_{f_1, \ldots, f_m}(V) \times A_{f_1, \ldots, f_m, 0, \ldots, 0}(W)). \tag{5.4.3}$$

Setting

$$\int_{U\bar{U}'=I} P'_f(U) \overline{P_g(U)} \dot{U} = R = R^{(q(f), q(g))}$$

and noting that under the transformations $V'UW$, where V and W are unitary matrices of order m and n, respectively, the integral remains unchanged, we obtain for the matrix R the relation

$$R = A'_1 R \bar{A}_2,$$

$$A_1 = A_{f_1, \ldots, f_m}(V) \times A_{f_1, \ldots, f_m, 0, \ldots, 0}(W),$$

$$A_2 = A_{g_1, \ldots, g_m}(V) \times A_{g_1, \ldots, g_m, 0, \ldots, 0}(W).$$

From this relation follows that for $f \neq g$, the matrix $R=0$, and for $f=g$, we have $R=\beta_f \cdot I$. In other words, the functions $\psi^{(i)}_{f_1,\dots,f_m}(U)$ form on $\mathfrak{C}_{\mathrm{I}}$ an orthogonal system. We have still to calculate the normalizing factors β_f,

$$\beta_f = \int_{U\bar{U}'=I} |\psi_f^{(i)}(U)|^2 \dot{U},$$

which, as can be seen from the condition $R=\beta_f \cdot I$, do not depend on f.

Since

$$q(f)\,\beta_f = \int_{U\bar{U}'=I} \sum_i |\psi_f^{(i)}(U)|^2 \dot{U}$$

$$= \int_{U\bar{U}'=I} \chi_f(U\bar{U}')\dot{U} = N(f_1, \dots, f_m) \int_{U\bar{U}'=I} \dot{U},$$

we have

$$\beta_f = \frac{c}{N(f_1, \dots, f_m, 0, \dots, 0)},$$

where

$$c = \frac{\omega_n}{\omega_{n-m}} = \frac{(2\pi)^{mn-\frac{m(m-1)}{2}}}{(n-m)! \dots (n-1)!}.$$

Setting $\rho=1$ in Theorem 1.2.5, we obtain the formula

$$\prod_{i=1}^{m} (1-x_i)^{-n} = \sum_{l_1 > \dots > l_m \geqslant 0} N(f_1, \dots, f_m, 0, \dots, 0)$$
$$\times \chi_{f_1,\dots,f_m}([x_1, \dots, x_m]) \quad (l_j = f_j + m - j).$$

For $Z \in \mathfrak{R}_{\mathrm{I}}$, $U \in \mathfrak{U}_{m,n}$, the moduli of the eigenvalues of the matrix $Z\bar{U}'$ are less than unity, so that

$$[\det(I - Z\bar{U}')]^{-n}$$
$$= \sum_{l_1 > \dots > l_m \geqslant 0} N(f_1, \dots, f_m, 0, \dots, 0)\,\chi_{f_1,\dots,f_m}(Z\bar{U}')$$
$$= c \sum_f \frac{1}{\beta_f} \sum_i \psi_f^{(i)}(Z)\overline{\psi_f^{(i)}(U)}. \tag{5.4.4}$$

Hence it readily follows that for $0<r<1$ the series (5.4.4) converges uniformly for Z lying in the region $rI - Z\bar{Z}' > 0$.

If $f(U)$ is a function which is integrable on $\mathfrak{U}_{m,n}$, then by multiplying (5.4.4) by $f(U)$ and integrating by parts we obtain

$$\int_{U\bar{U}'=I} [\det(I-Z\bar{U}')]^{-n} f(U)\,\dot{U} = \sum_{f,i} a_f^{(i)} \frac{\psi_f^{(i)}(Z)}{\sqrt{\beta_f}},$$

where

$$a_f^{(i)} = \frac{c}{\sqrt{\beta_f}} \cdot \int_{U\bar{U}'=I} f(U)\overline{\psi_f^{(i)}(U)} \cdot \dot{U}.$$

Thus we have arrived at the following theorem.

THEOREM 5.4.1. *Suppose that $f(U)$ is an integrable function, and let*

$$a_f^{(i)} = \frac{(2\pi)^{mn-\frac{m(m-1)}{2}}}{(n-m)!\ \dots\ (n-1)!} \int_{U\bar{U}'=I} f(U) \frac{\overline{\psi_f^{(i)}(U)}}{\sqrt{\beta_f}} \dot{U}$$

be the Fourier coefficients of this function with respect to the orthonormal system

$$\left\{\frac{1}{\sqrt{\beta_f}} \psi_f^{(i)}(U)\right\}.$$

Then the integral

$$\int_{U\bar{U}'=I} [\det(I-Z\bar{U}')]^{-n} f(U)\,\dot{U}$$

represents in the region $I-Z\bar{Z}'>0$ an analytic function which can be expanded in this region in the series

$$\sum_{f,i} a_f^{(i)} \frac{\psi_f^{(i)}(Z)}{\sqrt{\beta_f}}.$$

5.5. **Integrals of Cauchy type.** We shall confine ourselves to the case $m=n$, as the other cases are obtained in an entirely analogous way. We shall try to ascertain for which Z the integral

$$F(Z) = \int_U f(U)\{\det(I-Z\bar{U}')\}^{-n} \cdot \dot{U} \tag{5.5.1}$$

exists for a given integrable function $f(U)$.

THEOREM 5.5.1. *If the eigenvalues of the matrix $Z\bar{Z}'$ are either all larger or all smaller than unity, then for any unitary matrix U*

$$\det(I - Z\bar{U}') \neq 0.$$

In the opposite case one can always find a unitary matrix U such that

$$\det(I - Z\bar{U}') = 0.$$

PROOF. Suppose that $\det(I - Z\bar{U}') = 0$. Then it is possible to find a vector z, such that

$$z(I - Z\bar{U}') = 0,$$

i.e.,

$$z = zZ\bar{U}', \quad \bar{z}' = U\bar{Z}'\bar{z}',$$

hence

$$z\bar{z}' = zZ\bar{Z}'\bar{z}',$$

from which the first part of the theorem follows immediately.

In order to prove the second part of the theorem, let us note in advance that, without loss of generality, we can assume that

$$Z = [\lambda_1, \lambda_2, \ldots, \lambda_n], \quad \lambda_\nu \geqslant 0.$$

From the expression

$$\det\left[I - \begin{pmatrix} \lambda_1 & 0 \\ 0 & \lambda_2 \end{pmatrix}\begin{pmatrix} \cos\theta & \sin\theta \\ -\sin\theta & \cos\theta \end{pmatrix}\right] = \begin{vmatrix} 1 - \lambda_1\cos\theta & -\lambda_1\sin\theta \\ \lambda_2\sin\theta & 1 - \lambda_2\cos\theta \end{vmatrix}$$
$$= 1 - (\lambda_1 + \lambda_2)\cos\theta + \lambda_1\lambda_2 = 0$$

we obtain

$$\cos\theta = \frac{1 + \lambda_1\lambda_2}{\lambda_1 + \lambda_2}. \tag{5.5.2}$$

If $\lambda_1 \geqq 1 \geqq \lambda_2$, then $1 + \lambda_1\lambda_2 \leqq \lambda_1 + \lambda_2$, so that a θ which satisfies condition (5.5.2) exists, whereby the second part of the theorem is also proved.

THEOREM 5.5.2. *The integral* (5.5.1) *is meaningful in either of the two regions $I - Z\bar{Z}' > 0$ and $I - Z\bar{Z}' < 0$.*

Let us now proceed to a more detailed study of integrals of Cauchy type. We shall first of all extend the definition of the functions $\psi^{(i)}_{f_1,\ldots,f_n}(U)$ to the case of f_i's of any sign. We shall do this by means of the formula

$$\psi^{(i)}_{f_1,\ldots,f_n}(U) = \psi^{(i)}_{f_1 - f_n, f_2 - f_n, \ldots, f_{n-1} - f_n, 0}(U) \cdot (\det U)^{f_n},$$

for any $f_1 \geqq f_2 \geqq \cdots \geqq f_n$. Suppose now that the Fourier series for $f(U)$ has the form

$$\sum_{f_1 \geqslant f_2 \geqslant \cdots \geqslant f_n} \sum_i a^{(i)}_{f_1, \ldots, f_n} \cdot (\beta_{f_1, \ldots, f_n})^{-\frac{1}{2}} \cdot \psi^{(i)}_{f_1, \ldots, f_n}(U).$$

Then for Z belonging to the region $I - Z\overline{Z}' > 0$ we have

$$F(Z) = \sum_{f_1 \geqslant \cdots \geqslant f_n \geqslant 0} \sum_i a^{(i)}_{f_1, \ldots, f_n} (\beta_{f_1, \ldots, f_n})^{-\frac{1}{2}} \psi^{(i)}_{f_1, \ldots, f_n}(Z). \quad (5.5.3)$$

For Z belonging to the region $I - Z\overline{Z}' < 0$ we have

$$F(Z) = \int_{U\overline{U}' \leqslant 1} f(U)[\det(I - Z\overline{U}')]^{-n} \cdot \dot{U}$$

$$= [-\det Z]^{-n} \int_U f(U)(\det U)^n \cdot [\det(I - UZ^{-1})]^{-n} \cdot \dot{U}.$$

Since

$$(\det U)^n [\det(I - UZ^{-1})]^{-n}$$

$$= c(\det U)^n \sum_{f_1 \geqslant \cdots \geqslant f_n \geqslant 0} (\beta_{f_1, \ldots, f_n})^{-1} \sum_i \psi^{(i)}_{f_1, \ldots, f_n}(U) \cdot \psi^{(i)}_{f_1, \ldots, f_n}(Z^{-1})$$

$$= c \sum_{f_1 \geqslant \cdots \geqslant f_n \geqslant 0} (\beta_{f_1, \ldots, f_n})^{-1} \sum_i \psi^{(i)}_{f_1+n, \ldots, f_n+n}(U) \psi^{(i)}_{f_1, \ldots, f_n}(Z^{-1}),$$

it follows that by multiplying by $f(U)$ and then integrating by parts, we obtain

$$F(Z) = \int_{U\overline{U}' = 1} f(U)[\det(I - Z\overline{U}')]^{-n} \cdot \dot{U}$$

$$= (-\det Z)^{-n} \sum_{f_1 \geqslant \cdots \geqslant f_n \geqslant 0} \sum_i b^{(i)}_{f_1, \ldots, f_n} (\beta_{f_1, \ldots, f_n})^{-\frac{1}{2}} \psi^{(i)}_{f_1, \ldots, f_n}(Z^{-1}),$$

where

$$b^{(i)}_{f_1, \ldots, f_n} = c(\beta_{f_1, \ldots, f_n})^{-\frac{1}{2}} \int_{U\overline{U}' = 1} f(U) \psi^{(i)}_{f_1+n, \ldots, f_n+n}(U) \dot{U}$$

$$= c(\beta_{f_1, \ldots, f_n})^{-\frac{1}{2}} \int_{U\overline{U}' = 1} f(U) \overline{\psi^{(i)}_{-f_n-n, -f_{n-1}-n, \ldots, -f_1-n}(U)} \cdot \dot{U}$$

$$= a^{(i)}_{-f_n-n,\,\dots,\,-f_1-n} \quad 1).$$ [15]

Hence in the region $I - Z\bar{Z}' < 0$ the integral (5.5.1) represents an analytic function of Z^{-1} which has the expansion

$$F(Z) = \sum_{f_1 > \dots \geqslant f_n \geqslant 0} \sum_i a^{(i)}_{-f_n-n,\,\dots,\,-f_1-n} \left(\beta_{f_1,\dots,f_n}\right)^{-\frac{1}{2}} \psi^{(i)}_{-f_n-n,\dots,-f_1-n}(Z^{-1}). \tag{5.5.4}$$

5.6. **Differential operators.**[16] First let us introduce the differential operator

$$\partial_Z = \begin{pmatrix} \dfrac{\partial}{\partial z_{11}} & \cdots & \dfrac{\partial}{\partial z_{1n}} \\ \cdot & \cdot\cdot\cdot\cdot\cdot & \cdot \\ \dfrac{\partial}{\partial z_{m1}} & \cdots & \dfrac{\partial}{\partial z_{mn}} \end{pmatrix}. \tag{5.6.1}$$

Suppose Γ_{I} is a group of motions of the domain $\mathfrak{R}_{\mathrm{I}}$. It is well known (see L. K. Hua [1]) that Γ_{I} consists of transformations of the form

$$W = (AZ + B)(CZ + D)^{-1} = (Z\bar{B}' + \bar{A}')^{-1}(Z\bar{D}' + \bar{C}'), \tag{5.6.2}$$

where A is an $m \times m$ matrix, B is an $m \times n$ matrix, C is an $n \times m$ matrix,

---

[15] Translator's note. The author uses here the relation $\psi^{(i)}_{f_1,\dots,f_n}(U) = \psi^{(i)}_{-f_n,\dots,-f_1}(\bar{U})$. Let us show how this relation is obtained.

We shall consider the irreducible representation $A_{f_1,\dots,f_n}(U)$. Then $\overline{A_{-f_n,\dots,-f_1}(U)} = A_{-f_n,\dots,-f_1}(\bar{U})$ is also an irreducible representation; it is required to show that it is equivalent to the original representation. For this purpose we shall use the concept of highest weight of an irreducible representation.

Suppose $U = [e^{i\varphi_1},\dots,e^{i\varphi_n}]$ is a diagonal matrix. Then $A_{f_1,\dots,f_n}(U)$ is also (in an appropriate basis) a diagonal matrix, with numbers of the form $e^{i(m_1\varphi_1+\dots+m_n\varphi_n)}$ as the diagonal elements. The systems of integers $(m_1,\dots,m_n)$ are called the weights of the representation $A_{f_1,\dots,f_n}(U)$. Let us order the weights lexicographically. Then, as can easily be shown, the highest weight of the representation $A_{f_1,\dots,f_n}(U)$ is precisely the system $(f_1,\dots,f_n)$, and the lowest is the system $(f_n,\dots,f_1)$. It is well known that the representations $A_{f_1,\dots,f_n}(U)$ exhaust, to within equivalence, all the irreducible representations of the group $\mathfrak{U}$ of unitary matrices. Hence each irreducible representation of the group $\mathfrak{U}$ is completely determined by its highest (or lowest) weight.

Let us now consider the representation $\overline{A_{-f_n,\dots,-f_1}(U)}$. It is evident that the weights of the representations $\overline{A_{-f_n,\dots,-f_1}(U)}$ and $A_{-f_n,\dots,-f_1}(U)$ differ only by their sign. As the lowest weight of the representation $A_{-f_n,\dots,-f_1}(U)$ is $(-f_1,\dots,-f_n)$, the highest weight of the representation $\overline{A_{-f_n,\dots,-f_1}(U)}$ will be $(f_1,\dots,f_n)$. Hence the representations $\overline{A_{-f_n,\dots,-f_1}(U)}$ and $A_{f_1,\dots,f_n}(U)$ are equivalent. This completes the proof.

[16] Hua, L. K. and Look, K. H. [**3**].

and D is an $n\times n$ matrix.

These matrices satisfy the conditions

$$A\bar{A}'-B\bar{B}'=I,\quad A\bar{C}'=B\bar{D}',\quad C\bar{C}'-D\bar{D}'=-I, \tag{5.6.3}$$

which can also be written in the form

$$\bar{A}'A-\bar{C}'C=I,\quad \bar{A}'B=\bar{C}'D,\quad \bar{B}'B-\bar{D}'D=-I. \tag{5.6.4}$$

By differentiation of (5.6.2) we obtain, from (5.6.4),

$$\begin{aligned} dW &= [A-(AZ+B)(CZ+D)^{-1}C]\,dZ\cdot(CZ+D)^{-1}\\ &= [A-(Z\bar{B}'+\bar{A}')^{-1}(Z\bar{D}'+\bar{C}')C]\,dZ\cdot(CZ+D)^{-1}\\ &= (Z\bar{B}'+\bar{A}')^{-1}\,dZ\cdot(CZ+D)^{-1}. \end{aligned} \tag{5.6.5}$$

In contravariant notation this becomes

$$\partial'_W=(CZ+D)\,\partial'_Z\,(Z\bar{B}'+\bar{A}'). \tag{5.6.6}$$

Setting

$$W_1=(AZ_1+B)(CZ_1+D)^{-1}, \tag{5.6.7}$$

we have

$$\begin{aligned} I-\bar{W}_1'W &= I-(\bar{Z}_1'\bar{C}'+\bar{D}')^{-1}(\bar{Z}_1'\bar{A}'+\bar{B}')(AZ+B)(CZ+D)^{-1}\\ &= (\bar{Z}_1'\bar{C}'+\bar{D}')^{-1}(I-\bar{Z}_1'Z)(CZ+D)^{-1} \end{aligned} \tag{5.6.8}$$

and analogously

$$I-W\bar{W}_1'=(Z\bar{B}'+\bar{A}')^{-1}(I-Z\bar{Z}_1')(B\bar{Z}_1'+A)^{-1}. \tag{5.6.9}$$

By denoting

$$\Delta_Z=(I-Z\bar{Z}')\bar{\partial}_Z\cdot(I-\bar{Z}'Z)\cdot\partial'_Z,$$

we hence obtain

$$\Delta_W=(Z\bar{B}'+\bar{A}')^{-1}\Delta_Z\cdot(Z\bar{B}'+\bar{A}'). \tag{5.6.10}$$

The operator

$$\mathrm{Sp}\Delta_Z$$

we shall call the Laplace operator of the domain $\mathfrak{R}_{\mathrm{I}}$. [Let us note that $\bar{\partial}_Z\cdot(I-\bar{Z}'Z)$ signifies only formal matrix-multiplication, i.e., the elements of the matrix $I-\bar{Z}'Z$ are not differentiated.] In more expanded notation the operator $\mathrm{Sp}\Delta_Z$ has the following form:

$$\operatorname{Sp}\Delta_Z=\sum_{i,j=1}^{m}\sum_{\beta,\gamma=1}^{n}\left(\delta_{ij}-\sum_{\alpha=1}^{n}z_{i\alpha}\bar{z}_{j\alpha}\right)\left(\delta_{\beta\gamma}-\sum_{k=1}^{n}\bar{z}_{k\beta}z_{k\gamma}\right)\frac{\partial^2}{\partial z_{i\gamma}\partial\bar{z}_{j\beta}}. \quad (5.6.11)$$

This expression for the Laplace operator is invariant under transformations of the form (5.6.2), i.e., with respect to the group of motions of the domain $\mathfrak{R}_{\mathrm{I}}$.

DEFINITION 1. A real function $u(Z)$ which satisfies the equation $\operatorname{Sp}\Delta_Z\cdot u=0$, we shall call harmonic in $\mathfrak{R}_{\mathrm{I}}$.

5.7. **The meaning of the Laplace operator on the boundary of $\mathfrak{R}_{\mathrm{I}}$.** Let $\overline{\mathfrak{R}}_{\mathrm{I}}$ denote the closure of $\mathfrak{R}_{\mathrm{I}}$. The set of points of $\overline{\mathfrak{R}}_{\mathrm{I}}$, such that the matrix $I-Z\bar{Z}'$ has rank r, we shall denote by $\mathfrak{C}^{(r)}$, $r=0,1,\cdots,m$. It is clear that $\overline{\mathfrak{R}}_{\mathrm{I}}$ is equal to the sum of all $\mathfrak{C}^{(r)}$, $r=0,1,\cdots,m$, and that $\mathfrak{C}^{(0)}=\mathfrak{C}_{\mathrm{I}}$, $\mathfrak{C}^{(m)}=\mathfrak{R}_{\mathrm{I}}$.

DEFINITION 2. Suppose U and V are two fixed unitary matrices of order m and n, respectively. The point set

$$Z=U\begin{pmatrix} I^{(m-r)} & 0 \\ 0 & Z_0^{(r,\,n-m+r)}\end{pmatrix}V, \quad I-Z_0\bar{Z}_0'>0, \qquad (5.7.1)$$

is called an r-covering.

It is evident that each point of $\mathfrak{C}^{(r)}$ is contained in some r-covering, but two different coverings can have common points.

Let us now establish the meaning of the Laplace operator on the boundary of $\mathfrak{R}_{\mathrm{I}}$. Suppose Z is a point on $\mathfrak{C}^{(r)}$, and (5.7.1) is any of its r-coverings. As the Laplace operator is invariant under the transformations $Z_1=UZV$, it is sufficient to consider r-coverings of the form

$$Z=\begin{pmatrix} I^{(m-r)} & 0 \\ 0 & Z_0^{(r,\,n-m+r)}\end{pmatrix}, \quad I-Z_0\bar{Z}_0'>0. \qquad (5.7.2)$$

Let the function $u(Z)$ be defined in $\overline{\mathfrak{R}}_{\mathrm{I}}-\mathfrak{C}_{\mathrm{I}}$ and have continuous second-order derivatives with respect to all elements of Z on any r-covering, $r>0$. For the points of the r-covering (5.7.2) the operator $\operatorname{Sp}\Delta_Z$ reduces to the form

$$\operatorname{Sp}\Delta_Z=\operatorname{Sp}\left\{(I-Z_0\bar{Z}_0')\,\bar{\partial}_{Z_0}\cdot(I-\bar{Z}_0'Z_0)\cdot\partial_{Z_0}'\right\}.$$

We shall assume that

$$(\operatorname{Sp}\Delta_Z)\,u(Z)=\operatorname{Sp}\left\{(I-Z_0\bar{Z}_0')\,\bar{\partial}_{Z_0}\cdot(I-\bar{Z}_0'Z_0)\,\partial_{Z_0}'\right\}u\begin{pmatrix}1 & 0\\ 0 & Z_0\end{pmatrix}. \quad (5.7.3)$$

DEFINITION 3. A real function $u(Z)$ with the above

differential properties is called harmonic in $\overline{\mathfrak{R}}_{\mathrm{I}}$ if it satisfies the condition

$$(\mathrm{Sp}\,\Delta_Z)\,u(Z)=0 \tag{5.7.4}$$

at every point of $\overline{\mathfrak{R}}_{\mathrm{I}}-\mathfrak{C}_{\mathrm{I}}$.

It is clear that the harmonic property is invariant with respect to the group Γ_{I}. The set of harmonic functions is linear.

DEFINITION 4. The totality of harmonic functions in $\overline{\mathfrak{R}}_{\mathrm{I}}$ which are continuous on $\mathfrak{C}_{\mathrm{I}}$ we shall call class $\mathfrak{H}$.

THEOREM 5.7.1. *For any function $\varphi(U)$ which is continuous on $\mathfrak{C}_{\mathrm{I}}$ the Poisson integral*

$$u(Z)=\int_{\mathfrak{C}_I}\varphi(U)\,P_{\mathrm{I}}(Z,\ U)\,\dot{U} \tag{5.7.5}$$

represents a harmonic function of class $\mathfrak{H}$.

In order to prove this theorem, it is convenient to use two auxiliary propositions.

THEOREM 5.7.2. *If U on $\mathfrak{C}_{\mathrm{I}}$ is mapped into V (also on $\mathfrak{C}_{\mathrm{I}}$) by the transformation* (5.6.2), *then*

$$P_{\mathrm{I}}(W,\ V)=P_{\mathrm{I}}(Z,\ U)\,|\det(B\bar{U}'+A)|^{2n}. \tag{5.7.6}$$

PROOF. From (5.6.8) and (5.6.9) follows that

$$\begin{aligned}&(I-W\bar{V}')^{-1}(I-W\bar{W}')(I-V\bar{W}')^{-1}\\&=(B\bar{U}'+A)(I-Z\bar{U}')^{-1}(I-Z\bar{Z}')(I-U\bar{Z}')^{-1}(U\bar{B}'+\bar{A}').\end{aligned} \tag{5.7.7}$$

Hence

$$\frac{[\det(I-W\bar{W}')]^n}{|\det(I-W\bar{V}')|^{2n}}=\frac{[\det(I-Z\bar{Z}')]^n}{|\det(I-Z\bar{U}')|^{2n}}|\det(B\bar{U}'+A)|^{2n},$$

whereby Theorem 5.7.2 is proved.

THEOREM 5.7.3. *The Poisson kernel $P_{\mathrm{I}}(Z,U)$ (with fixed U) is a harmonic function in $\mathfrak{R}_{\mathrm{I}}$ but does not belong to the class $\mathfrak{H}$.*

PROOF. We shall first show that

$$(\mathrm{Sp}\,\Delta_Z)\,P_{\mathrm{I}}(Z,\ U)=0$$

for $Z=0$. We shall set

$$Z=(z_{j\alpha}),\quad U=(u_{j\alpha}),\quad 1\leqslant j\leqslant m,\quad 1\leqslant \alpha\leqslant n.$$

Hence we obtain

$$(\mathrm{Sp}\,\Delta_Z)P_{\mathrm{I}}(Z,U)|_{Z=0}=\mathrm{Sp}\left(\bar{\partial}_Z\cdot\partial_Z'\right)\cdot P_{\mathrm{I}}(Z,U)|_{Z=0}$$

$$=\left[\sum_{j=1}^{m}\sum_{\alpha=1}^{n}\frac{\partial^2}{\partial\bar{z}_{j\alpha}\,\partial z_{j\alpha}}P_{\mathrm{I}}(Z,U)\right]_{Z=0}$$

$$=\frac{1}{V(\mathfrak{C}_{\mathrm{I}})}\cdot\sum_{j=1}^{m}\sum_{\alpha=1}^{n}\frac{\partial^2}{\partial\bar{z}_{j\alpha}\,\partial z_{j\alpha}}\frac{[\det(I-Z\bar{Z}')]^n}{[\det(I-Z\bar{U}')(I-U\bar{Z}')]^n}\bigg|_{Z=0}$$

$$=\frac{1}{V(\mathfrak{C}_{\mathrm{I}})}\sum_{j=1}^{m}\sum_{\alpha=1}^{n}\left\{\frac{\partial^2}{\partial\bar{z}_{j\alpha}\partial z_{j\alpha}}[\det(I-Z\bar{Z}')]^n\right.$$

$$\left.+\frac{\partial}{\partial z_{j\alpha}}\det(I-Z\bar{U}')^{-n}\frac{\partial}{\partial\bar{z}_{j\alpha}}[\det(I-\bar{Z}'U)]^{-n}\right\}\bigg|_{Z=0}$$

$$=\frac{1}{V(\mathfrak{C}_{\mathrm{I}})}\sum_{j=1}^{m}\sum_{\alpha=1}^{n}\left\{\frac{\partial^2}{\partial\bar{z}_{j\alpha}\partial z_{j\alpha}}\left(1-n\sum_{\beta=1}^{n}\sum_{k=1}^{m}|z_{k\beta}|^2\right)\right.$$

$$\left.+\frac{\partial}{\partial z_{j\alpha}}\left(1+n\sum_{\beta=1}^{n}\sum_{k=1}^{m}z_{k\beta}\bar{u}_{k\beta}\right)\frac{\partial}{\partial\bar{z}_{j\alpha}}\left(1+n\sum_{\beta=1}^{n}\sum_{k=1}^{m}\bar{z}_{k\beta}u_{k\beta}\right)\right\}\bigg|_{Z=0}=0.$$

From the transitivity of the domain $\mathfrak{R}_{\mathrm{I}}$ with respect to the group Γ_{I}, and from Theorem 5.7.2, we obtain the assertion of the theorem for any interior point of the domain $\mathfrak{R}_{\mathrm{I}}$. Theorem 5.7.3 is proved.

By differentiating under the integral sign, we obtain for any interior point of $\mathfrak{R}_{\mathrm{I}}$

$$(\mathrm{Sp}\,\Delta_Z)\,u(Z)=0.$$

However, the proof of Theorem 5.7.1 is not yet complete. We have to ascertain the behavior of the integral (5.7.5) on the boundary of $\mathfrak{R}_{\mathrm{I}}$.

5.8. **The behavior of the Poisson integral on the boundary of** $\mathfrak{R}_{\mathrm{I}}$. In order to complete the proof of Theorem 5.7.1, it is sufficient to demonstrate the following two theorems.

THEOREM 5.8.1. *For any* $V\in\mathfrak{C}_{\mathrm{I}}$

$$\lim_{Z\to V} u(Z)=\varphi(V). \tag{5.8.1}$$

THEOREM 5.8.2. *Suppose $m>1$, and*

$Q=\begin{pmatrix}1 & 0\\ 0 & Z_0\end{pmatrix}$, *where Z_0 is an $(m-1)\times(n-1)$ matrix,*

$$I^{(m-1)}-Z_0\bar{Z}_0'>0. \tag{5.8.2}$$

Then

$$\lim_{Z\to Q}\int_{\mathfrak{C}_{\mathrm{I}}}\varphi(U)P_{\mathrm{I}}(Z,U)\dot{U}=\int_{\mathfrak{C}_{\mathrm{I}}^{m-1,\,n-1}}\varphi\begin{pmatrix}1 & 0\\ 0 & U_0\end{pmatrix}P_{\mathrm{I}}^{m-1,\,n-1}(Z_0,U_0)\dot{U}_0, \tag{5.8.3}$$

where $\mathfrak{C}_{\mathrm{I}}^{m-1,n-1}$ and $\mathfrak{R}_{\mathrm{I}}^{m-1,n-1}$ denote $\mathfrak{C}_{\mathrm{I}}$ and $\mathfrak{R}_{\mathrm{I}}$ for $(m-1)\times(n-1)$ matrices, and $P_{\mathrm{I}}^{m-1,n-1}(Z_0,U_0)$ denotes the Poisson kernel for the domain $\mathfrak{R}_{\mathrm{I}}^{m-1,n-1}$.

From Theorem 5.8.2 it follows that it is sufficient to prove Theorem 5.8.1 for $m=1$. We shall put Theorem 5.8.1 in a somewhat different form for this case, by slightly altering the notations.

THEOREM 5.8.1′. *We shall set $v=(v_1,v_2,\cdots,v_n)$. For any v which satisfies the condition $v\bar{v}'=1$, and for any function $\varphi(u)$ which is continuous on the manifold $u\bar{u}'=1$, we have*

$$\varphi(v)=\lim_{z\to v}\frac{1}{V(\mathfrak{C}_{\mathrm{I}})}\int_{u\bar{u}'=1}\varphi(u)\frac{(1-z\bar{z}')^n}{|1-z\bar{u}'|^{2n}}\dot{u}. \tag{5.8.4}$$

PROOF. As the vector v can be mapped by a unitary transformation into the vector $e=(1,0,\cdots,0)$, we can assume that $v=e$.

First let us prove (5.8.4) for $z=\rho e$, $\rho\to 1$, $0<\rho<1$, i.e.,

$$\varphi(e)=\lim_{\rho\to 1}\frac{1}{V(\mathfrak{C}_{\mathrm{I}})}\int_{u\bar{u}'=1}\varphi(u)\frac{(1-\rho^2)^n}{|1-\rho\bar{u}_1|^{2n}}\dot{u}. \tag{5.8.5}$$

For $\varphi(u)\equiv 1$, formula (5.8.5) holds, so that

$$1=\frac{1}{V(\mathfrak{C}_{\mathrm{I}})}\int_{u\bar{u}'=1}\frac{(1-\rho^2)^n}{|1-\rho\bar{u}_1|^{2n}}\dot{u}. \tag{5.8.6}$$

Hence it is enough to prove that for any positive ϵ one can take ρ_0 so close to unity that for $\rho_0\leqq\rho<1$

$$\left| \int\limits_{u\bar{u}'=1} (\varphi(u) - \varphi(e)) \frac{(1-\rho^2)^n}{|1-\rho\bar{u}_1|^{2n}} \dot{u} \right| < \varepsilon. \tag{5.8.7}$$

We shall use the parametric representation of the manifold $u\bar{u}'=1$:

$u\bar{u}'=1$:

$$u_1 = \cos\theta_1 + i\sin\theta_1 \cdot \cos\theta_2$$

$$u_2 = \sin\theta_1 \cdot \sin\theta_2 \cdot \cos\theta_3 + i\sin\theta_1 \cdot \sin\theta_2 \cdot \sin\theta_3 \cdot \cos\theta_4,$$

$$\cdots\cdots\cdots\cdots\cdots\cdots\cdots$$

$$u_{n-1} = \sin\theta_1 \cdot \ldots \cdot \sin\theta_{2n-4} \cdot \cos\theta_{2n-3} + i\sin\theta_1 \cdot \ldots \cdot \sin\theta_{2n-4} \cdot \sin\theta_{2n-3} \cdot \cos\theta_{2n-2},$$

$$u_n = \sin\theta_1 \cdot \ldots \cdot \sin\theta_{2n-2} \cdot \cos\theta_{2n-1} + i\sin\theta_1 \cdot \ldots \cdot \sin\theta_{2n-2} \cdot \sin\theta_{2n-1},$$

$$0 \leqslant \theta_j < \pi, \quad j = 1, 2, \ldots, 2n-2; \quad 0 \leqslant \theta_{2n-1} < 2\pi.$$

For any positive ϵ one can take δ so small that

$$|\varphi(u) - \varphi(e)| < \frac{\varepsilon}{2}$$

for $0 \leqq \theta_1 \leqq \delta$. Hence

$$\frac{1}{V(\mathfrak{C}_{\mathrm{I}})} \left| \int\limits_{\substack{u\bar{u}'=1 \\ 0\leqslant\theta_1\leqslant\delta}} (\varphi(u) - \varphi(e)) \frac{(1-\rho^2)^n}{|1-\rho\bar{u}_1|^{2n}} \cdot \dot{u} \right| < \frac{\varepsilon}{2}.$$

Further, with the already chosen δ we have for $\delta < \theta_1 < \pi$

$$|1-\rho\bar{u}_1|^{2n} = \{(1-\rho\cos\theta_1)^2 + \rho^2\sin^2\theta_1\cos^2\theta_2\}^n$$

$$\geqslant (1-\rho\cos\theta_1)^{2n} \geqslant (1-\cos\delta)^{2n} = 2^{2n}\sin^{4n}\frac{\delta}{2}.$$

We shall take ρ so close to unity that

$$(1-\rho)^n < \frac{2^{n-1}\sin^{4n}\frac{\delta}{2}}{M} \cdot \frac{\varepsilon}{2}, \qquad M = \sup_{u\bar{u}'=1} |\varphi(u)|.$$

Then

$$\frac{1}{V(\mathfrak{C}_{\mathrm{I}})} \left| \int\limits_{\substack{u\bar{u}'=1 \\ \delta\leqslant\theta_1\leqslant\pi}} (\varphi(u) - \varphi(e)) \frac{(1-\rho^2)^n}{|1-\rho\bar{u}_1|^{2n}} \dot{u} \right|$$

$$\leqslant \frac{2M \cdot 2^n}{V(\mathfrak{C}_{\mathrm{I}})} \cdot \frac{(1-\rho)^n}{2^{2n}\sin^{4n}\frac{\delta}{2}} \cdot \int\limits_{u\bar{u}'=1} \dot{u} \leqslant \frac{\varepsilon}{2}.$$

Thus formula (5.8.7) is proved.

Hence (5.8.4) holds for $z \to v$ along the radius. But since this approach to the limit is uniform and $\varphi(u)$ is continuous, it follows that (5.8.4) holds also for $z \to v$ along any path.

Before proceeding to the proof of Theorem 5.8.2, let us make a remark. From Poisson's formula (5.8.7) evidently follows the formula

$$u(0) = \frac{1}{V(\mathfrak{C}_{\mathrm{I}})} \cdot \int_{\mathfrak{C}_{\mathrm{I}}} \varphi(U) \cdot \dot{U} \tag{5.8.8}$$

(the mean-value theorem). However, by using transformation (5.6.2) we can without difficulty obtain (5.7.5) from (5.8.8).

Proof of Theorem 5.8.2. By virtue of the above remark it is sufficient to prove that

$$\lim_{Z \to \begin{pmatrix} 1 & 0 \\ 0 & 0 \end{pmatrix}} \int_{\mathfrak{C}_{\mathrm{I}}} \varphi(U) P_{\mathrm{I}}(Z, U) \dot{U} = \int_{\mathfrak{C}_{\mathrm{I}}^{(m-1, n-1)}} \varphi\left[\begin{pmatrix} 1 & 0 \\ 0 & U_0 \end{pmatrix}\right] \cdot \dot{U}_0. \tag{5.8.9}$$

Let us first consider the case when

$$Z = \begin{pmatrix} \rho & 0 \\ 0 & 0 \end{pmatrix}, \quad \rho \to 1, \quad 0 < \rho < 1.$$

Writing

$$U = \begin{pmatrix} u \\ U_1 \end{pmatrix}, \quad u = (u_1, \ldots, u_n),$$

we have

$$\int_{\mathfrak{C}_{\mathrm{I}}} \varphi(U) P_{\mathrm{I}}(Z, U) \dot{U} = \frac{1}{V(\mathfrak{C}_{\mathrm{I}})} \int_{\mathfrak{C}_{\mathrm{I}}} \varphi(U) \frac{(1-\rho^2)^n}{|1-\rho \bar{u}_1|^{2n}} \dot{U}$$

$$= \frac{1}{V(\mathfrak{C}_{\mathrm{I}})} \int_{u\bar{u}'=1} \frac{(1-\rho^2)^n}{|1-\rho \bar{u}_1|^{2n}} \tau(u) \dot{u}, \tag{5.8.10}$$

where

$$\tau(u) = \int_{U_1} \varphi(U) \dot{U}_1.$$

Since $\tau(u)$ is continuous on the manifold $u\bar{u}'=1$ and

$$V(\mathfrak{C}_{\mathrm{I}}) = V\left(\mathfrak{C}_{\mathrm{I}}^{(m-1, n-1)}\right) \cdot \int_{u\bar{u}'=1} \dot{u},$$

we obtain from Theorem 5.8.1′

$$\int_{\mathfrak{C}_{\mathrm{I}}} \varphi(U) P_{\mathrm{I}}(Z, U) \dot{U} = \frac{1}{V\left(\mathfrak{C}_{\mathrm{I}}^{(m-1, n-1)}\right)} \tau(e), \tag{5.8.11}$$

where $e = (1, 0, \cdots, 0)$. Since $U\bar{U}' = I$, we have (for $u = e$):

$$U_1 = (0, U_0), \quad U_0 = U_0^{(m-1, n-1)} \text{ and } \tau(e) = \int_{U_0} \varphi\begin{pmatrix} 1 & 0 \\ 0 & U_0 \end{pmatrix} \cdot \dot{U}_0.$$

Thus equality (5.8.3) is proven for the case that Z approaches Q along a fixed path. But again, from the uniformity of this approach it follows that (5.8.3) is also valid when Z approaches Q along any path.

Hence we have proved Theorem 5.8.2, and thereby also Theorem 5.8.1.

5.9. **The solution of Dirichlet's problem in $\mathfrak{R}_{\mathrm{I}}$.** In §§5.7 and 5.8 we have shown that for any function $\varphi(U)$ which is continuous on $\mathfrak{C}_{\mathrm{I}}$, the Poisson integral (5.7.5) gives us a harmonic function of class $\mathfrak{H}$, with $\varphi(U)$ as its boundary values.

Let us now proceed to the proof of the uniqueness of the solution of this problem, setting out from the maximum principle.

THEOREM 5.9.1. *A harmonic function of class $\mathfrak{H}$ attains its maximum and minimum on the manifold $\mathfrak{C}_{\mathrm{I}}$.*

Let us note that it is enough to prove this theorem when $\mathfrak{C}_{\mathrm{I}}$ has been replaced by the entire boundary of $\mathfrak{R}_{\mathrm{I}}$, since the problem which arises on each $\mathfrak{C}^{(r)}$, $0<r<m$, is analogous (for $\mathfrak{R}_{\mathrm{I}}^{(r, n-m+r)}$).

The proof is preceded by the following theorem.

THEOREM 5.9.2. *Suppose that $\rho(Z)$ is a real function and that $v(Z)$ is the solution of the equation in partial derivatives*

$$(\mathrm{Sp}\, \Delta_Z)\, v(Z) = \rho(Z). \tag{5.9.1}$$

If $\rho(Z) > 0$, then $v(Z)$ cannot attain a maximum in $\mathfrak{R}_{\mathrm{I}}$, and if $\rho(Z) < 0$, it cannot attain a minimum.

PROOF. Since the second part of the theorem is obtained at once from the first part by replacing $\rho(Z)$ and $v(Z)$ by $-\rho(Z)$ and $-v(Z)$, we can confine ourselves to the proof of the first part.

Let us assume that $v(Z)$ attains its maximum at an interior point Z_0, which we can take as zero without loss of generality. Then (5.9.1) yields

$$\sum_{j=1}^{m}\sum_{\alpha=1}^{n}\frac{\partial^2}{\partial z_{j\alpha}\partial \bar{z}_{j\alpha}}v(Z)\bigg|_{Z=0}=\rho(0)>0. \tag{5.9.2}$$

But since $v(Z)$ has a maximum at the point $Z=0$, it follows that

$$\frac{\partial^2}{\partial z_{j\alpha}\partial \bar{z}_{j\alpha}}v(Z)\bigg|_{Z=0}\leqslant 0,$$

which contradicts (5.9.2), and the theorem is proved.

PROOF OF THEOREM 5.9.1. Let us denote by M the exact upper limit of $u(Z)$ on the boundary of $\mathfrak{R}_{\mathrm{I}}$. Suppose that there exists in the interior of $\mathfrak{R}_{\mathrm{I}}$ a point W_0 such that

$$u(W_0)>M+\varepsilon. \tag{5.9.3}$$

Let us construct the auxiliary function

$$v(Z)=u(Z)+\eta\,\mathrm{Sp}\,[(Z-W_0)\overline{(Z-W_0)}'],$$

where η has been chosen so small that

$$\eta\sup_{Z\in\overline{\mathfrak{R}}_{\mathrm{I}}}\mathrm{Sp}\,[(Z-W_0)\overline{(Z-W_0)}']<\frac{\varepsilon}{2}.$$

For any point P on the boundary of $\mathfrak{R}_{\mathrm{I}}$, we have

$$\begin{aligned}v(W_0)&=u(W_0)\geqslant u(P)+\varepsilon\\&=v(P)-\eta\,\mathrm{Sp}\,((P-W_0)\overline{(P-W_0)}')+\varepsilon>v(P)+\frac{\varepsilon}{2}.\end{aligned}$$

Hence $v(Z)$ attains its maximum at an interior point of $\mathfrak{R}_{\mathrm{I}}$. But

$$\begin{aligned}(\mathrm{Sp}\,\Delta_Z)\,v(Z)&=(\mathrm{Sp}\,\Delta_Z)\,u(Z)+\eta\,(\mathrm{Sp}\,\Delta_Z)\,\{\mathrm{Sp}\,[(Z-W_0)\overline{(Z-W_0)}']\}\\&=\eta\,\mathrm{Sp}\,(I-Z\bar{Z}')\,\mathrm{Sp}\,(I-\bar{Z}'Z).\end{aligned} \tag{5.9.4}$$

Since $\mathrm{Sp}(I-Z\bar{Z}')>0$ and $\mathrm{Sp}(I-\bar{Z}'Z)>0$, we arrive at a contradiction to Theorem 5.9.2.

Theorem 5.9.1 leads immediately to important consequences.

THEOREM 5.9.3. *The only harmonic function of class $\mathfrak{H}$ which has zero boundary-values on $\mathfrak{C}_{\mathrm{I}}$ is the identically vanishing function.*

THEOREM 5.9.4. *Each harmonic function of class $\mathfrak{H}$ is uniquely determined by its boundary values on $\mathfrak{C}_{\mathrm{I}}$ by means of Poisson's formula*

$$u(Z)=\frac{1}{V(\mathfrak{C}_{\mathrm{I}})}\int_{\mathfrak{C}_{\mathrm{I}}}u(U)\,P(Z,U)\,\dot{U}.$$

REMARK. One can show directly that

$$\Delta_Z P(Z, U) = 0.$$

This is a matrix equation, equivalent to m^2 second-order differential equations. Hence any harmonic function in $\mathfrak{R}_{\mathrm{I}}$ satisfies m^2 second-order differential equations.

5.10. **A basis for harmonic functions.** Suppose now $m=n$, and that

$$A_{f_1, \dots, f_n}(U) = A_f(U) = \left(a^f_{ij}(U)\right)_{i,j=1}^{N(f)}$$

denotes an irreducible unitary representation of the unitary group $\mathfrak{U}_n$ with signature $f=(f_1,\cdots,f_n)$; as we already know, the order of this representation is $N(f)=N(f_1,\cdots,f_n)$. It is easy to see that the set of functions

$$\varphi^f_{ij}(U) = \sqrt{\frac{N(f)}{V(\mathfrak{C}_{\mathrm{I}})}}\, a^f_{ij}(U) \tag{5.10.1}$$

forms an orthonormal system on $\mathfrak{C}_{\mathrm{I}}$. By Theorem 1.2.5, with $\rho=1$,

$$[\det(I - Z\overline{U}')]^{-n} = \sum_{f \geqslant 0} N(f)\, \mathrm{Sp}\,[A_f(Z\overline{U}')]$$

$$= V(\mathfrak{C}_{\mathrm{I}}) \sum_{f_1 \geqslant \dots \geqslant f_n \geqslant 0} \; \sum_{i,j=1}^{N(f)} \varphi^f_{ij}(Z)\, \overline{\varphi^f_{ij}(U)}, \tag{5.10.2}$$

so that the series converges uniformly for Z lying in the closure of the region

$$rI - Z\overline{Z}' > 0, \quad 0 < r < 1. \tag{5.10.3}$$

The functions

$$a^f_{ij}(U)\, \overline{a^g_{st}(U)}, \quad f \geqslant 0, \quad g \geqslant 0,$$

are the elements of the matrix

$$A_f(U) \times \overline{A_g(U)},$$

which can be decomposed into the direct sum

$$\sum_h A_h(U),$$

where $N(f)N(g) = \sum_h N(h)$, $h=(h_1, h_2, \cdots, h_n)$, $h_1 \geqq h_2 \geqq \cdots \geqq h_n$.[17]

---

[17] Translator's note. In other words, the space of the representation $A_f(U) \times \overline{A_g(U)}$ is decomposed into the direct sum of irreducible invariant subspaces. $A_h(U)$ is the matrix of the corresponding irreducible representations.

Consequently

$$a_{ij}^{f}(U)\,\overline{a_{st}^{g}(U)}=\sum \lambda_{pq}^{h} a_{pq}^{h}(U)$$

and $\sum |\lambda_{pq}^{h}|^2=1$. Hence we obtain

$$\varphi_{ij}^{f}(U)\,\overline{\varphi_{st}^{g}(U)}=\sum \mu_{pq}^{h}\varphi_{pq}^{h}(U), \tag{5.10.4}$$

where

$$\mu_{pq}^{h}=\lambda_{pq}^{h}\sqrt{\frac{N(f)\,N(g)}{V(\mathfrak{C}_{\mathrm{I}})\,N(h)}}.$$

From (5.10.2) we obtain the expansion of the Poisson kernel

$$\frac{1}{V(\mathfrak{C}_{\mathrm{I}})}\frac{[\det(I-Z\bar{Z}')]^n}{|\det(I-Z\bar{U}')|^{2n}}$$

$$=V(\mathfrak{C}_{\mathrm{I}})\cdot\sum_{f\geqslant 0}\sum_{g\geqslant 0}\{\det(I-Z\bar{Z}')\}^n\sum_{i,j}\sum_{s,t}\overline{\varphi_{ij}^{f}(U)}\,\varphi_{ij}^{f}(Z)\,\varphi_{st}^{g}(U)\,\varphi_{st}^{g}(\bar{Z})$$

$$=\sum_{h}\sum_{i,j}\Phi_{ij}^{h}(Z)\,\overline{\varphi_{ij}^{h}(U)}. \tag{5.10.5}$$

One can show that this series converges uniformly for Z lying in the closure of the set (5.10.3). Hence

$$\Phi_{ij}^{h}(Z)=\int_{\mathfrak{C}_{\mathrm{I}}} P_{\mathrm{I}}(Z,\ U)\,\varphi_{ij}^{h}(U)\,\dot{U},$$

$$\lim_{Z\to U}\Phi_{ij}^{h}(Z)=\varphi_{ij}^{h}(U).$$

Therefore we have the theorem:

THEOREM 5.10.1. *The system of harmonic functions*

$$\{\Phi_{ij}^{h}(Z)\}$$

forms a basis for the harmonic functions of class $\mathfrak{H}$.

5.11. Abelian summability of Fourier series on the unitary group.

DEFINITION. If

$$\lim_{Z\to U}\sum_{h}\sum_{i,j} c_{ij}^{h}\Phi_{ij}^{h}(Z)=\varphi(U),$$

we shall say that the series

$$\sum_h \sum_{i,j} c_{ij}^h \varphi_{ij}^h (U)$$

is A-summable or Abel-summable to the sum $\varphi(U)$.

THEOREM 5.11.1. *The Fourier series of a function which is continuous on* $\mathfrak{C}_{\mathrm{I}}$ *is A-summable to this function.*

This theorem is a direct consequence of Theorem 5.8.1 and of formula (5.10.5).

Let us now set

$$\mathfrak{B}_h(Z) = \sqrt{\frac{V(\mathfrak{C}_{\mathrm{I}})}{N(h)}} \left(\Phi_{ij}^h(Z)\right)_{i,j=1}^{N(h)}.$$

Then

$$\mathfrak{B}_h(Z) = \int_{\mathfrak{C}_{\mathrm{I}}} P_{\mathrm{I}}(Z, U) A_h(U) \dot{U}. \tag{5.11.1}$$

Any matrix Z of order n of the set (5.10.3) can be represented in the form

$$Z = V\Lambda W,$$

where V and W are unitary matrices, and $\Lambda = [\lambda_1, \cdots, \lambda_n]$, $0 \leqq \lambda_n \leqq \cdots \leqq \lambda_1 \leqq \sqrt{r}$. Substituting in (5.11.1), we obtain

$$\begin{aligned} \mathfrak{B}_h(Z) &= \int_{\mathfrak{C}_{\mathrm{I}}} P_{\mathrm{I}}(V\Lambda W, U) A_h(U) \dot{U} \\ &= \int_{\mathfrak{C}_{\mathrm{I}}} P_{\mathrm{I}}(\Lambda, \bar{V}' U \bar{W}') A_h(U) \dot{U} = A_h(V) \mathfrak{B}_h(\Lambda) A_h(W) \end{aligned} \tag{5.11.2}$$

(by replacing the variable U by VUW).

In particular, for $\Lambda = \lambda I$ we have

$$A_h(V) \mathfrak{B}_h(\lambda I) = \mathfrak{B}_h(\lambda I) A_h(V),$$

i.e., the matrix $\mathfrak{B}_h(\lambda I)$ commutes with the matrices $A_h(V)$ of the irreducible representation. By Schur's lemma

$$\mathfrak{B}_h(\lambda I) = \rho_h(\lambda) I^{(N(h))},$$

where $\rho(\lambda) \to 1$, when $\lambda \to 1$.

More precisely, we have the following particular case of the preceding theorem.

THEOREM 5.11.2. *For any function* $\varphi(U)$ *which is continuous on* $\mathfrak{C}_{\text{I}}$, *we have*

$$\varphi(U) = \lim_{\lambda \to 1} \sum_{h} \sum_{i,j} c_{ij}^{h} \rho_h(\lambda) \varphi_{ij}^{h}(U), \tag{5.11.3}$$

where

$$c_{ij}^{h} = \frac{1}{V(\mathfrak{C}_{\text{I}})} \int_{\mathfrak{C}_{\text{I}}} \varphi(U) \overline{\varphi_{ij}^{h}(U)}\, \dot{U}. \tag{5.11.4}$$

Hence we obtain at once the more precise form of the Peter-Weyl theorem.

THEOREM 5.11.3. *Any function* $\varphi(U)$ *which is continuous on the unitary group can be approximated as closely as desired by functions of the system* $\{\varphi_{ij}^{h}(U)\}$.

PROOF. From (5.11.3) it is evident that for any positive ϵ one can choose λ so close to unity that

$$\left| \varphi(U) - \sum_{h} \sum_{ij} c_{ij}^{h} \rho_h(\lambda) \varphi_{ij}^{h}(U) \right| < \frac{\varepsilon}{2},$$

and thereupon one can choose N so large that

$$\left| \varphi(U) - \sum_{N \geqslant h_1 \geqslant \cdots \geqslant h_n \geqslant -N} \sum_{i,j} c_{ij}^{h} \rho_h(\lambda) \varphi_{ij}^{h}(U) \right| < \varepsilon.$$

REMARK 1. Since any continuous function on the compact subgroup of the unitary group $\mathfrak{U}_n$ can be extended in a continuous way to the entire group $\mathfrak{U}_n$, Theorem 5.11.3 yields, as a corollary, the corresponding result for any compact group of finite dimensionality (this involving some orthonormalization of the system $\{\varphi_{ij}^{h}(U)\}$ on the subgroup).

REMARK 2. Any continuous function on the set of cosets of $\mathfrak{U}_n$ with respect to any compact subgroup can be considered as a continuous function on $\mathfrak{U}_n$. Therefore we obtain, as a corollary of Theorem 5.11.3, the corresponding result for compact homogeneous finite-dimensional spaces [with corresponding averaging of $\varphi_{ij}^{h}(U)$ with respect to the subgroup].

CHAPTER VI

HARMONIC ANALYSIS IN THE SPACE OF SYMMETRIC AND SKEW-SYMMETRIC MATRICES

6.1. **Orthonormal systems in the space of symmetric unitary matrices.** We shall consider orthonormal systems in the spaces $\mathfrak{R}_{\text{II}}$ and $\mathfrak{C}_{\text{II}}$. The transformation

$$T = USU' \tag{6.1.1}$$

with unitary matrix U maps the symmetric unitary matrix S into the symmetric unitary matrix T. We shall arrange the elements of the matrix S as a vector s

$$\big(s_{11}, \sqrt{2}s_{12}, \sqrt{2}s_{13}, \ldots, \sqrt{2}s_{1n}, s_{22}, \sqrt{2}s_{23}, \ldots$$
$$\ldots, \sqrt{2}s_{2n}, \ldots, s_{n-1,n-1}, \sqrt{2}s_{n-1,n}, s_{nn}\big)$$

of dimension $n(n+1)/2$, and we shall carry out a similar operation for the matrix T, denoting the vector so obtained by t. When the matrix S is mapped into the matrix T by the transformation (6.1.1), the vector s is mapped into the vector t by a linear transformation with matrix $U^{[2]}$ of order $n(n+1)/2$.

From the vector s we shall construct the vector $s^{[f]}$ of dimension $\frac{(n(n+1)/2+f-1)!}{f!\,(n(n+1)/2-1)!}$.

The matrix of the transformation of $s^{[f]}$ into $t^{[f]}$ is

$$\big(U^{[2]}\big)^{[f]}. \tag{6.1.2}$$

The components of $s^{[f]}$ are monomials of s_{ij} of degree f. These components are linearly independent, and any homogeneous polynomial of degree f can be written in the form of a linear combination of the components.

From Theorems 1.3.2 and 1.4.2 we know that the space of homogeneous polynomials of degree f of s_{ij} can be decomposed into a direct sum of subspaces which are invariant under the transformation (6.1.2). These subspaces have dimension $N(2f_1, \cdots, 2f_n)$, where $f_1+\cdots+f_n=f$. The polynomials which form the basis of an $N(2f_1, \cdots, 2f_n)$-dimensional subspace we shall denote by

$$\varphi^{(i)}_{f_1, \cdots, f_n}(S), \qquad i=1, 2, \ldots, N(2f_1, \ldots, 2f_n). \tag{6.1.3}$$

When S is mapped into T by the transformation (6.1.1), $\varphi^{(i)}_{f_1,\cdots,f_n}(S)$ is mapped into $\varphi^{(i)}_{f_1,\cdots,f_n}(T)$ by the transformation with matrix $A_{2f_1,\cdots,2f_n}(U)$. Moreover, system (6.1.3) is orthogonal on $\mathfrak{C}_{\mathrm{II}}$, i.e.,

$$\int_{\mathfrak{C}_{\mathrm{II}}} \varphi^{(i)}_{f_1, \ldots, f_n}(S)\, \overline{\varphi^{(j)}_{g_1, \ldots, g_n}(S)}\, \dot{S} = \delta_{ij}\delta_{fg} \cdot \rho_f,$$

where $\dot{S}$ is the volume element of the space of symmetric unitary matrices, found in §3.5.

By setting

$$\psi^{(i)}_{f_1, \ldots, f_n}(S) = \rho^{-\frac{1}{2}}_{f_1, \ldots, f_n} \cdot \varphi^{(i)}_{f_1, \ldots, f_n}(S),$$

we obtain an orthonormal system in $\mathfrak{C}_{\mathrm{II}}$.

6.2. **Projection of a kernel into a subspace.** Let us consider the functions

$$\Psi_{f_1, \ldots, f_n}(S, \overline{T}) = \sum_i \psi^{(i)}_{f_1, \ldots, f_n}(S)\, \overline{\psi^{(i)}_{f_1, \ldots, f_n}(T)}. \tag{6.2.1}$$

We directly obtain

$$\int_{\mathfrak{C}_{\mathrm{II}}} \Psi_{f_1, \ldots, f_n}(S, \overline{S})\, \dot{S} = N(2f_1, \ldots, 2f_n). \tag{6.2.2}$$

Since the $\Psi_{f_1,\cdots,f_n}(S,\overline{S})$ do not depend on S, it follows that

$$\Psi_{f_1, \ldots, f_n}(S, \overline{S}) = \frac{1}{V(\mathfrak{C}_{\mathrm{II}})} \cdot N(2f_1, \ldots, 2f_n).$$

Thus $\Psi_{f_1,\cdots,f_n}(S,\overline{S})$ are very simple in form. We shall try to also simplify expression (6.2.1).

Let us order lexicographically the index system $(f_1, \cdots, f_n)$. We shall write $f>g$, where $f=(f_1, \cdots, f_n)$, $g=(g_1, \cdots, g_n)$, if $f_1=g_1$, $f_2=g_2, \cdots, f_{q-1}=g_{q-1}$, $f_q>g_q$.

From (6.1.1) follows

$$A_{f_1, \cdots, f_n}(T) = A_{f_1, \cdots, f_n}(U)\, A_{f_1, \cdots, f_n}(S)\, A_{f_1, \cdots, f_n}(U'). \tag{6.2.3}$$

Suppose L is a linear space, generated by the elements of the matrices $A_{f_1,\cdots,f_n}(S)$. When S is mapped into T, L is mapped into itself. This means that L is an invariant subspace. This invariant subspace can be decomposed into a direct sum of invariant subspaces, corresponding to $A_{2g_1,\cdots,2g_n}(X)$. Evidently, $(2g_1,\cdots,2g_n)\leqq(2f_1,\cdots,2f_n)$, whereby the equality certainly obtains. The expression

$$\mathrm{Sp}\left[A_{f_1,\ldots,f_n}(S)\,A_{f_1,\ldots,f_n}(\bar{T})\right]=\mathrm{Sp}\,A_{f_1,\ldots,f_n}(S\bar{T})$$

is invariant on replacing S and T simultaneously by USU' and UTU'. Since

$$\mathrm{Sp}\,A_{f_1,\ldots,f_n}(S\bar{T})=\chi_{f_1,\ldots,f_n}(S\bar{T}),$$

it follows that

$$\chi_{f_1,\ldots,f_n}(S\bar{T})=\sum_{g\leqslant f}c_{f,g}\,\Psi_{g_1,\ldots,g_n}(S,\,\bar{T}), \tag{6.2.4}$$

whereby $c_{f,f}\neq 0$. This relation is reversible, i.e., we can write

$$\Psi_{f_1,\ldots,f_n}(S,\,\bar{T})=\sum_{g\leqslant f}d_{f,g}\,\chi_{g_1,\ldots,g_n}(S\bar{T}). \tag{6.2.5}$$

From the last formula it is evident that

$$\Psi_{f_1,\ldots,f_n}(S,\,\bar{T})=\Psi_{f_1,\ldots,f_n}(S\bar{T}).$$

We shall now endeavor to find an effective method of determining the coefficients of formula (6.2.5). Since $c_{f,g}$ and $d_{f,g}$ are determined only for $g\leqq f$, we shall set $c_{f,g}=d_{f,g}=0$ for $g>f$. Moreover, since we have ordered the index systems $f=(f_1,\cdots,f_n)$, we can now consider f as a simple index, i.e., $f=1,2,3,\cdots$. Then the matrices $C=(f_{f,g})_1^k$ and $D=(d_{f,g})_1^k$ will be nonsingular triangular matrices for any integral positive k.

Let us consider the integral

$$\int_S\int_T \Psi_f(S\bar{T})\,\chi_g(T\bar{S})\,\dot{S}\dot{T}, \qquad g\leqslant f.$$

Substituting in this integral the expression for χ_g given by formula (6.2.4), and using (6.2.1), we obtain

$$\sum_{h\leqslant g}c_{g,h}\int_S\int_T \Psi_f(S\bar{T})\,\Psi_h(T\bar{S})\,\dot{S}\dot{T}$$

$$=\sum_{h\leqslant g}c_{g,h}\sum_i\sum_j\int_S\int_T \psi_f^{(i)}(S)\,\overline{\psi_f^{(i)}(T)}\,\psi_h^{(j)}(T)\,\overline{\psi_h^{(j)}(S)}\,\dot{S}\dot{T}=\sum_{i,\,j}\sum_{h\leqslant g}c_{g,h}\delta_{ij}\delta_{fh},$$

whence follows

$$\int_S \int_T \Psi_f(S\bar{T})\chi_g(T\bar{S})\,\dot{S}\dot{T}=0 \quad \text{for} \quad g<f, \tag{6.2.6}$$

and

$$\int_S \int_T \Psi_f(S\bar{T})\chi_f(T\bar{S})\,\dot{S}\dot{T}=N_f c_{f,f}\neq 0 \ ^{1)}. \tag{6.2.7}$$

[18]

We shall set

$$\beta_{f,g}=\int_S \int_T \chi_f(S\bar{T})\chi_g(T\bar{S})\,\dot{S}\dot{T}. \tag{6.2.8}$$

From (6.2.5), (6.2.6), (6.2.7) and (6.2.8) we obtain

$$\sum_{h\leqq g} d_{f,h}\beta_{h,g}=0, \qquad g<f, \tag{6.2.9}$$

and

$$\sum_{h\leqq f} d_{f,h}\beta_{h,f}=N_f c_{f,f}\neq 0. \tag{6.2.10}$$

It is evident that the matrix $B=(\beta_{h,f})_1^k$ is positive definite for any integral $k\geqq 1$. Equations (6.2.9) and (6.2.10) can be replaced by the matrix equation

$$DB=\begin{pmatrix} N_1c_{1,1} & * & \cdots & * \\ 0 & N_2c_{2,2} & \cdots & * \\ \cdot & \cdot & \cdots & \cdot \\ 0 & 0 & \cdots & N_kc_{k,k} \end{pmatrix}=K. \tag{6.2.11}$$

Hence it is evident that the $d_{f,g}$ can be expressed by $\beta_{f,g}$ and by the elements of the matrix K. Since, however, the elements of the matrix K are unknown, we shall obtain from (6.2.10) another system of equations from which the $d_{f,g}$ are obtained recursively in terms of $\beta_{f,g}$ and some other quantities which can be readily calculated.

We shall put

$$\alpha_f=\int_S \chi_f(S\bar{S})\,\dot{S}. \tag{6.2.12}$$

Setting $T=S$ in (6.2.5) and integrating with respect to S, we obtain

$$\sum_{h\leqq f} d_{f,h}\,\alpha_h=N_f. \tag{6.2.13}$$

---

[18] Translator's note. $N_f=N(2f_1,\cdots,2f_n)$.

For $f=1$, we evidently have $d_{1,1}=N_1/c_{1,1}$. We shall assume that for $f\leqq k-1$ the quantities $d_{f,g}$ $(g=1,2,\cdots,f)$ are expressed by $\beta_{f,g}$ and α_f, and then investigate what happens for $f=k$. From (6.2.9) and (6.2.13) we find

$$\begin{pmatrix} d_{1,1} & 0 & \cdots & 0 \\ d_{2,1} & d_{2,2} & \cdots & 0 \\ \cdot & \cdot & \cdot & \cdot \\ d_{k,1} & d_{k,2} & \cdots & d_{k,k} \end{pmatrix} \begin{pmatrix} \beta_{1,1} & \beta_{1,2} & \cdots & \beta_{1,k-1} & \alpha_1 \\ \beta_{2,1} & \beta_{2,2} & \cdots & \beta_{2,k-1} & \alpha_2 \\ \cdot & \cdot & \cdot & \cdot & \cdot \\ \beta_{k,1} & \beta_{k,2} & \cdots & \beta_{k,k-1} & \alpha_k \end{pmatrix}$$

$$= \begin{pmatrix} d_{1,1}\beta_{1,1} & * & \cdots & * \\ 0 & d_{2,1}\beta_{1,2}+d_{2,2}\beta_{2,2} & \cdots & * \\ \cdot & \cdot & \cdot & \cdot \\ 0 & 0 & \cdots & N_k \end{pmatrix}. \tag{6.2.14}$$

The triangular matrix in the right-hand side is nonsingular, since by (6.2.10) its diagonal elements are nonvanishing. Since the matrix D is the transpose of the matrix C and is therefore nonsingular, it follows that the matrix

$$B_k=\begin{pmatrix} \beta_{1,1} & \cdots & \beta_{1,k-1} & \alpha_1 \\ \cdot & \cdot & \cdot & \cdot \\ \beta_{k,1} & \cdots & \beta_{k,k-1} & \alpha_k \end{pmatrix}$$

is also nonsingular. From (6.2.14) we have

$$(d_{k,1},\ d_{k,2},\ \ldots,\ d_{k,k})=(0,\ 0,\ \ldots,\ 0,\ N_k)\,B_k^{-1}.$$

This proves that for any f the $d_{f,g}$ are expressed in terms of $\beta_{f,g}$ and α_f. It remains to set forth an effective method of calculation of $\beta_{f,g}$ and α_f.

Any symmetric unitary matrix S can be written in the form $S=UU'$, where U is a unitary matrix. Setting $T=UWU'$, we obtain from

$$\chi_f(S\overline{T})=\chi_f(U\overline{W}\overline{U}')=\chi_f(\overline{W}),$$

the formula

$$\beta_{f,g}=\int_{\mathfrak{C}_{II}}\dot{S}\int_{\mathfrak{C}_{II}}\chi_g(W)\overline{\chi_f(W)}\,\dot{W}.$$

By (3.5.19) we have

$$\beta_{f,g}=V(\mathfrak{C}_{II})\cdot C\int_{2\pi\geqslant\theta_1\geqslant\cdots\geqslant\theta_n\geqslant 0}\cdots\int\chi_g([e^{i\theta_1},\ \ldots,\ e^{i\theta_n}])$$

$$\times\overline{\chi_f([e^{i\theta_1},\ \ldots,\ e^{i\theta_n}])}\prod_{1\leqslant\nu<\mu\leqslant n}|e^{i\theta_\nu}-e^{i\theta_\mu}|\,d\theta_1\ldots d\theta_n, \tag{6.2.15}$$

where

$$C = 2^{\frac{n(n-1)}{2}} \int_{\{O\}} \{\dot{O}\}.$$

Since

$$e^{i\theta_\nu} - e^{i\theta_\mu} = 2i \sin \frac{\theta_\nu - \theta_\mu}{2} \cdot e^{\frac{i}{2}(\theta_\nu + \theta_\mu)}, \tag{6.2.16}$$

it follows that

$$\beta_{f,g} = V(\mathfrak{C}_{\text{II}}) \cdot C \cdot 2^{\frac{n(n-1)}{2}} \int_{2\pi \geqslant \theta_1 \geqslant \dots \geqslant \theta_n \geqslant 0} \cdots \int \chi_g([e^{i\theta_1}, \dots, e^{i\theta_n}])$$

$$\times \chi_f([e^{-i\theta_1}, \dots, e^{-i\theta_n}]) \prod_{1 \leqslant \nu < \mu \leqslant n} \sin \frac{\theta_\nu - \theta_\mu}{2} \cdot d\theta_1 \dots d\theta_n. \tag{6.2.17}$$

The following method of calculation of $\beta_{f,g}$ is somewhat more convenient. By (6.2.16) we have for $2\pi \geqq \theta_\nu \geqq \theta_\mu \geqq 0$

$$|e^{i\theta_\nu} - e^{i\theta_\mu}| = 2 \sin \frac{\theta_\nu - \theta_\mu}{2} = -i\,(e^{i\theta_\nu} - e^{i\theta_\mu})\, e^{-\frac{i}{2}(\theta_\nu + \theta_\mu)}$$

and

$$\prod_{1 \leqslant \nu < \mu \leqslant n} |e^{i\theta_\nu} - e^{i\theta_\mu}|$$

$$= (-i)^{\frac{n(n-1)}{2}} \prod_{1 \leqslant \nu < \mu \leqslant n} (e^{i\theta_\nu} - e^{i\theta_\mu})\, e^{-\frac{i}{2}(n-1)(\theta_1 + \dots + \theta_n)}.$$

Hence

$$\beta_{f,g} = V(\mathfrak{C}_{\text{II}}) \cdot C \cdot (-i)^{\frac{n(n-1)}{2}} \int_{2\pi \geqslant \theta_1 \geqslant \dots \geqslant \theta_n \geqslant 0} \cdots \int \chi_g([e^{i\theta_1}, \dots, e^{i\theta_n}])$$

$$\times \chi_f([e^{-i\theta_1}, \dots, e^{-i\theta_n}]) \prod_{1 \leqslant \nu < \mu \leqslant n} (e^{i\theta_\nu} - e^{i\theta_\mu})$$

$$\times e^{-\frac{i}{2}(n-1)(\theta_1 + \dots + \theta_n)}\, d\theta_1 \dots d\theta_n. \tag{6.2.18}$$

Since the expression

$$\chi_g([e^{i\theta_1}, \dots, e^{i\theta_n}]) \prod_{1 \leqslant \nu < \mu \leqslant n} (e^{i\theta_\nu} - e^{i\theta_\mu})$$

coincides with $M_g([e^{i\theta_1}, \cdots, e^{i\theta_n}])$, and $\chi_f([e^{-i\theta_1}, \cdots, e^{-i\theta_n}])$ can be expressed by elementary symmetric functions of $e^{i\theta_1}, \cdots, e^{i\theta_n}$, we can calculate $\beta_{f,g}$ for any particular case. However, it is not possible to obtain the value of $\beta_{f,g}$ in the general case.

6.3. **An orthonormal system on $\mathfrak{R}_{\mathrm{II}}$.** From the results of the foregoing section we know that the functions

$$\psi_f^{(i)}(Z)$$

form in $\mathfrak{R}_{\mathrm{II}}$ a complete orthogonal system. It remains to compute the integral

$$\rho_f = \int_{\mathfrak{R}_{\mathrm{II}}} |\psi_f^{(i)}(Z)|^2 \dot{Z}.$$

Since it is well known that ρ_f does not depend on i, it follows that

$$\rho_f \cdot N(2f_1, \ldots, 2f_n) = \int_{\mathfrak{R}_{\mathrm{II}}} \Psi_f(Z\bar{Z}) \dot{Z}.$$

By formula (6.2.5), the problem reduces to the calculation of the integral

$$\int_{\mathfrak{R}_{\mathrm{II}}} \chi_f(Z\bar{Z}) \dot{Z}.$$

By (3.5.2), we have for this integral the expression

$$2^n \int_{\{\mathfrak{U}_n\}} \{\dot{U}_n\} \underset{0 \leqslant \lambda_1 \leqslant \cdots \leqslant \lambda_n \leqslant 1}{\int \cdots \int} \prod_{j<k} |\lambda_j^2 - \lambda_k^2| \cdot \chi_f([\lambda_1^2, \ldots, \lambda_n^2]) \lambda_1 \ldots \lambda_n \, d\lambda_1 \ldots d\lambda_n$$

$$= (-1)^{\frac{n(n-1)}{2}} \omega_n \underset{0 \leqslant \lambda_1 \leqslant \cdots \leqslant \lambda_n \leqslant 1}{\int \cdots \int} \det |\lambda_i^{2l_j}|_{i,j=1}^n \cdot \lambda_1 \ldots \lambda_n \, d\lambda_1 \ldots d\lambda_n$$

$$= (-1)^{\frac{n(n-1)}{2}} \omega_n \int_0^1 \cdots \int_0^1 \sum_{i_1, \ldots, i_n} \delta_{i_1 \ldots i_n}^{1 \ldots n} r_1^{2l_{i_1}+1} r_2^{2l_{i_1}+2l_{i_2}+3} \ldots r_n^{2l_{i_1}+\cdots+2l_{i_n}+2n-1} dr_1 \ldots dr_n$$

$$= (-1)^{\frac{n(n-1)}{2}} 2^{-n} \omega_n \int_0^1 \cdots \int_0^1 \sum_{i_1, \ldots, i_n} \delta_{i_1 \cdots i_n}^{1 \cdots n} t_1^{l_{i_1}} t_2^{l_{i_1}+l_{i_2}+1}$$

$$\ldots t_n^{l_{i_1}+\cdots+l_{i_n}+n-1}\, dt_1 \ldots dt_n = (-1)^{\frac{n(n-1)}{2}} 2^{-n} \omega_n \sum_{i_1,\ldots,i_n} \delta^{1\ldots n}_{i_1\ldots i_n}$$

$$\times \frac{1}{(l_{i_1}+1)(l_{i_1}+l_{i_2}+2)\cdots(l_{i_1}+\cdots+l_{i_n}+n)} = \frac{\omega_n D(l_1,\ldots,l_n)}{\prod_{1\leqslant i<j\leqslant n}(l_i+l_j+2)}.$$

In the last transformation we used the following identity.

LEMMA.

$$\sum_{i_1,\ldots,i_n} \delta^{1\ldots n}_{i_1\ldots i_n} \frac{1}{l_{i_1}(l_{i_1}+l_{i_2})\cdots(l_{i_1}+\cdots+l_{i_n})} = \frac{(-1)^{\frac{n(n-1)}{2}} 2^n D(l_1,\ldots,l_n)}{\prod_{1\leqslant i<j\leqslant n}(l_i+l_j)}.$$

PROOF. For $n=2$, the identity is trivial. Let us assume that it holds for $n-1$. Then for n its left-hand side is transformed into the expression

$$\frac{2^{n-1}}{l_1+l_2+\cdots+l_n} \sum_{i=1}^{n} (-1)^{n-i} \frac{D(l_1,\ldots,l_{i-1},l_{i+1},\ldots,l_n)}{\prod_{1\leqslant k<j\leqslant n}(l_k+l_j)} \prod_{j=1}^{n}(l_i+l_j)$$

$$= \frac{2^{n-1}}{\sigma_1}(-1)^{n-1} \prod_{1\leqslant k<j\leqslant n}(l_j+l_k)^{-1} \sum_{i=1}^{n}(-1)^{i-1}$$

$$\times D(l_1,\ldots,l_{i-1},l_{i+1},\ldots,l_n)\left(l_i^n+\sigma_1 l_i^{n-1}+\cdots+\sigma_n\right),$$

where σ_ν is an elementary symmetric function of order ν of $l_1,\cdots,l_n$. Since

$$\sum_{i=1}^{n}(-1)^{i-1} D(l_1,\ldots,l_{i-1},l_{i+1},\ldots,l_n)\, l_i^n = \sigma_1 D(l_1,\ldots,l_n),$$

$$\sum_{i=1}^{n}(-1)^{i-1} D(l_1,\ldots,l_{i-1},l_{i+1},\ldots,l_n)\, l_i^{n-1} = D(l_1,\ldots,l_n),$$

and all the other sums vanish, the lemma is proved.

6.4. Characteristic manifold of the space of skew-symmetric matrices. Manifold (3.6.16) is the characteristic manifold $\mathfrak{C}_{\mathrm{III}}$ of the domain $\mathfrak{R}_{\mathrm{III}}$. Let us consider the transformation

$$L = UKU', \tag{6.4.1}$$

which maps the skew-symmetric matrix K into the skew-symmetric matrix L (U is a unitary matrix). Arranging the elements of the matrix K in the form of the row

$$(k_{1,2}, \ldots, k_{1,n}, k_{2,3}, \ldots, k_{2,n}, \ldots, k_{n-1,n}),$$

we shall denote by k the $(n(n-1)/2)$-dimensional vector so obtained and by l the vector which corresponds to the matrix L. When the matrix K is mapped into the matrix L by the transformation (6.4.1), the vector k is mapped into the vector l by a linear transformation with matrix $U^{(2)}$, (see §1.4).

Let us now consider the vector $k^{[f]}$ of dimension

$$\frac{(n(n-1)/2+f-1)!}{f!\,(n(n-1)/2-1)!}.$$

When the matrix K is mapped into L by the transformation (6.4.1), the vector $k^{[f]}$ is mapped into $l^{[f]}$ by a linear transformation with matrix

$$\left(U^{(2)}\right)^{[f]}. \tag{6.4.2}$$

From Theorems 1.4.3 and 1.3.3 we know that the space which spans the vectors $k^{[f]}$ can be decomposed into a direct sum of invariant subspaces of dimensionality $N(f_1, f_1, f_2, f_2, \cdots)$. The basis vectors (i.e., homogeneous polynomials of the elements of K) in these subspaces we shall denote by

$$\varphi^{(i)}_{f_1, \ldots, f_{[n/2]}}(K), \qquad i = 1, 2, \ldots, N(f_1, f_1, f_2, f_2, \ldots).$$

It can be readily shown that

$$\int_{\mathfrak{R}_{\mathrm{III}}} \varphi^{(i)}_f(Z)\overline{\varphi^{(j)}_g(Z)}\,\dot{Z} = \rho_f \cdot \delta_{ij}\delta_{fg},$$

and

$$\int_{\mathfrak{C}_{\mathrm{III}}} \varphi^{(i)}_f(K)\overline{\varphi^{(j)}_g(K)}\,\dot{K} = \beta_f\delta_{ij}\delta_{fg}.$$

Further results for skew-symmetric matrices are obtained in exactly the analogous way as for symmetric matrices.

CHAPTER VII

HARMONIC ANALYSIS ON LIE SPHERES

7.1. **Gegenbauer polynomials.** For the convenience of the reader we shall start with an exposition of some basic facts of the theory of Gegenbauer polynomials and of harmonic analysis on the ordinary Euclidean sphere (in n-dimensional space).

The Gegenbauer polynomials[19] are defined by the formula

$$C_m^\lambda(\xi) = \sum_{s=0}^{[m/2]} (-1)^s \frac{\Gamma(m+\lambda-s)}{s!\,(m-2s)!\,\Gamma(\lambda)} (2\xi)^{m-2s}, \qquad \lambda > -\frac{1}{2}. \tag{7.1.1}$$

It can be shown that

$$\sum_{m=0}^{\infty} C_m^\lambda(\xi)\, w^m = (1-2\xi w + w^2)^{-\lambda}. \tag{7.1.2}$$

We shall find also convenient the formula

$$(1-\xi^2)^{\lambda-\frac{1}{2}} C_m^\lambda(\xi) = \frac{(-2)^m}{m!} \frac{\Gamma(m+\lambda)\Gamma(m+2\lambda)}{\Gamma(\lambda)\Gamma(2m+2\lambda)} \left(\frac{d}{d\xi}\right)^m (1-\xi^2)^{m+\lambda-\frac{1}{2}}, \tag{7.1.3}$$

called Rodrigues' formula. This formula can be readily proved from (7.1.1), by showing that $(d/d\xi)\ C_m^\lambda(\xi) = 2\lambda C_{m-1}^{\lambda+1}(\xi)$ and using mathematical induction.

Suppose now $f(\xi)$ is a function which is m times differentiable on the interval $[-1,1]$. From (7.1.3) we have

$$\int_{-1}^{1} f(\xi)\, C_m^\lambda(\xi) \left(1-\xi^2\right)^{\lambda-\frac{1}{2}} d\xi$$

$$= \frac{(-2)^m}{m!} \cdot \frac{\Gamma(m+\lambda)\Gamma(m+2\lambda)}{\Gamma(\lambda)\Gamma(2m+2\lambda)} \int_{-1}^{1} f(\xi) \left(\frac{d}{d\xi}\right)^m (1-\xi^2)^{m+\lambda-\frac{1}{2}} d\xi. \tag{7.1.4}$$

---

[19] Translator's note. In the Chinese original these polynomials are called ultraspherical and are denoted by $P_m^{(\lambda)}$.

Integrating by parts, we obtain

$$\int_{-1}^{1} f(\xi)\left(\frac{d}{d\xi}\right)^{m}(1-\xi^2)^{m+\lambda-\frac{1}{2}}\,d\xi$$

$$=f(\xi)\left(\frac{d}{d\xi}\right)^{m-1}(1-\xi^2)^{m+\lambda-\frac{1}{2}}\Big|_{-1}^{1}-\int_{-1}^{1} f'(\xi)\left(\frac{d}{d\xi}\right)^{m-1}(1-\xi^2)^{m+\lambda-\frac{1}{2}}\,d\xi.$$

Since $\lambda>-\frac{1}{2}$, it follows that $(d/d\xi)^{m-1}(1-\xi^2)^{m+\lambda-1/2}|_{-1}^{1}=0$, and hence

$$\int_{-1}^{1} f(\xi)\left(\frac{d}{d\xi}\right)^{m}(1-\xi^2)^{m+\lambda-\frac{1}{2}}\,d\xi=-\int_{-1}^{1} f'(\xi)\left(\frac{d}{d\xi}\right)^{m-1}(1-\xi^2)^{m+\lambda-\frac{1}{2}}\,d\xi.$$

Repeating this operation, we finally obtain

$$\int_{-1}^{1} f(\xi)\,C_m^{\lambda}(\xi)\left(1-\xi^2\right)^{\lambda-\frac{1}{2}}\,d\xi$$

$$=\frac{2^m}{m!}\cdot\frac{\Gamma(m+\lambda)\,\Gamma(m+2\lambda)}{\Gamma(\lambda)\,\Gamma(2m+2\lambda)}\int_{-1}^{1}(1-\xi^2)^{m+\lambda-\frac{1}{2}}\left(\frac{d}{d\xi}\right)^{m} f(\xi)\,d\xi. \qquad (7.1.5)$$

If $f(\xi)$ is a polynomial of degree m with leading coefficient a, then

$$\int_{-1}^{1} f(\xi)\,C_m^{\lambda}(\xi)\left(1-\xi^2\right)^{\lambda-\frac{1}{2}}d\xi=2^m\frac{\Gamma(m+\lambda)\,\Gamma(m+2\lambda)}{\Gamma(\lambda)\,\Gamma(2m+2\lambda)}\int_{-1}^{1} a(1-\xi^2)^{m+\lambda-\frac{1}{2}}d\xi$$

$$=2^m\cdot\frac{\Gamma(m+\lambda)\,\Gamma(m+2\lambda)\,\Gamma\left(m+\lambda+\frac{1}{2}\right)\Gamma\left(\frac{1}{2}\right)}{\Gamma(\lambda)\,\Gamma(2m+2\lambda)\,\Gamma(m+\lambda+1)}\,a. \qquad (7.1.6)$$

Setting in particular $f(\xi)=C_l^{\lambda}(\xi)$, we obtain from (7.1.6) and (7.1.5)

$$\int_{-1}^{1} C_l^{\lambda}(\xi)\,C_m^{\lambda}(\xi)\left(1-\xi^2\right)^{\lambda-\frac{1}{2}}d\xi=\begin{cases}0, & l\neq m,\\ \dfrac{\pi\cdot 2^{1-2\lambda}\cdot\Gamma(m+2\lambda)}{[\Gamma(\lambda)]^2(m+\lambda)\,m!}, & l=m.\end{cases} \qquad (7.1.7)$$

If we now set in formula (7.1.5) $f(\xi)=\xi^l$, then for $l\geqq m$

$$\int_{-1}^{1}\xi^l C_m^{\lambda}(\xi)\left(1-\xi^2\right)^{\lambda-\frac{1}{2}}d\xi$$

$$=\binom{l}{m}\frac{2^m\,\Gamma(m+\lambda)\,\Gamma(m+2\lambda)}{\Gamma(\lambda)\,\Gamma(2m+2\lambda)}\int_{-1}^{1}\xi^{l-m}\left(1-\xi^2\right)^{m+\lambda-\frac{1}{2}}d\xi. \qquad (7.1.8)$$

If $l-m$ is an odd number, this integral vanishes, and if $l-m=2k$, then from the formula

$$\int_{-1}^{1} \xi^{l-m}\left(1-\xi^{2}\right)^{m+\lambda-\frac{1}{2}} d\xi$$

$$=\int_{0}^{1} \xi^{k-\frac{1}{2}}(1-\xi)^{m+\lambda-\frac{1}{2}} d\xi=\frac{\Gamma\left(k+\frac{1}{2}\right)\Gamma\left(m+\lambda+\frac{1}{2}\right)}{\Gamma(k+m+\lambda+1)}$$

we obtain

$$\int_{-1}^{1} \xi^{l} C_{m}^{\lambda}(\xi)\left(1-\xi^{2}\right)^{\lambda-\frac{1}{2}} d\xi$$

$$=\begin{cases} 0, & \text{for} \quad l<m \quad \text{or} \quad l=m+2k+1, \\ \dfrac{\pi}{2^{l+2\lambda-1}} \cdot \dfrac{l!}{k!(l-2k)!} \dfrac{\Gamma(l-2k+2\lambda)}{\Gamma(\lambda)\Gamma(l-k+\lambda+1)}, & l=m+2k. \end{cases} \tag{7.1.9}$$

Any polynomial $f(\xi)$ of degree m can be represented in the form

$$f(\xi)=\sum_{l=0}^{m} a_{l} C_{l}^{\lambda}(\xi).$$

Multiplying this expression by $C_{l}^{\lambda}(\xi)(1-\xi^{2})^{\lambda-1/2}$ and integrating, we obtain

$$a_{l}=\frac{2^{2\lambda-1}[\Gamma(\lambda)]^{2}(l+\lambda)\cdot l!}{\pi\Gamma(l+2\lambda)} \int_{-1}^{1} f(\xi) C_{l}^{\lambda}(\xi)\left(1-\xi^{2}\right)^{\lambda-\frac{1}{2}} d\xi.$$

From this we easily derive the following theorem.

THEOREM 7.1.1. *If $f(\xi)$ is a polynomial of degree m for which the integrals in the last formula vanish for $0 \leq l \leq m-1$, then $f(\xi)$ differs from $C_{m}^{\lambda}(\xi)$ by a constant factor only.*

From this theorem and from (7.1.9) we easily obtain

$$\xi^{m}=\frac{m!\,\Gamma(\lambda)}{2^{m}} \sum_{k=0}^{[m/2]} \frac{m-2k+\lambda}{k!\,\Gamma(m-k+\lambda+1)} C_{m-2k}^{\lambda}(\xi). \tag{7.1.10}$$

In the following we shall also find useful the formula: for $\nu>\lambda$

$$C_{m}^{\nu}(\xi)=\frac{\Gamma(\lambda)}{\Gamma(\nu)} \cdot \sum_{k=0}^{[m/2]} c_{k} C_{m-2k}^{\lambda}(\xi), \tag{7.1.11}$$

where

$$c_k = \frac{m-2k+\lambda}{k!}\cdot\frac{\Gamma(k+\nu-\lambda)}{\Gamma(\nu-\lambda)}\cdot\frac{\Gamma(m+\nu-k)}{\Gamma(m+\lambda-k+1)}. \tag{7.1.12}$$

This formula can be proved as follows. From (7.1.1) and (7.1.10) we have

$$C_m^\nu(\xi) = \sum_{s=0}^{[m/2]}(-1)^s\frac{\Gamma(\nu+m-s)}{\Gamma(\nu)\Gamma(s+1)\Gamma(m-2s+1)}(2\xi)^{m-2s}$$

$$=\frac{\Gamma(\lambda)}{\Gamma(\nu)}\sum_{s=0}^{[m/2]}(-1)^s\frac{\Gamma(\nu+m-s)}{\Gamma(s+1)}\sum_{k=0}^{[m/2]-s}\frac{m-2s-2k+\lambda}{k!\,\Gamma(m-2s-k+\lambda+1)}C^\lambda_{m-2s-2k}(\xi)$$

$$=\frac{\Gamma(\lambda)}{\Gamma(\nu)}\sum_{t=0}^{[m/2]}c_tC^\lambda_{m-2t}(\xi),$$

where

$$c_t=\sum_{s+k=t}(-1)^s\frac{\Gamma(\nu+m-s)(m-2s-2k+\lambda)}{s!\,k!\,\Gamma(m-2s-k+\lambda+1)}$$

$$=\frac{m-2t+\lambda}{t!}\sum_{s=0}^{t}(-1)^s\binom{t}{s}\frac{\Gamma(\nu+m-s)}{\Gamma(m-t-s+\lambda+1)}$$

$$=\frac{(m-2t+\lambda)\Gamma(t+\nu-\lambda)\Gamma(\nu+m-t)}{t!\,\Gamma(\nu-\lambda)\Gamma(m-t+\lambda+1)}.$$

[Here we have made use of the formula for the qth difference

$$\Delta^q f(x)=\sum_{s=0}^{q}(-1)^s\binom{q}{s}f(x+q-s).$$

Since $\Delta(\Gamma(\alpha+x)/\Gamma(\beta+x))=(\alpha-\beta)(\Gamma(\alpha+x)/\Gamma(\beta+x+1))$, it follows that for $\alpha>\beta$

$$\Delta^q\frac{\Gamma(\alpha+x)}{\Gamma(\beta+x)}=\frac{\Gamma(\alpha-\beta+1)}{\Gamma(\alpha-\beta-q+1)}\cdot\frac{\Gamma(\alpha+x)}{\Gamma(\beta+x+q)}.]$$

7.2. **Harmonic analysis on the sphere.** Suppose γ is the unit sphere in real n-dimensional Euclidean space, i.e., the locus of the points $x=(x_1,\cdots,x_n)$ which satisfy the relation

$$xx'=1. \tag{7.2.1}$$

It is well known that the unit sphere admits the following parametric representation:

$$x_\nu = \sin\theta_1 \cdot \ldots \cdot \sin\theta_{\nu-1} \cdot \cos\theta_\nu, \qquad 1 \leqslant \nu \leqslant n-1,$$
$$x_n = \sin\theta_1 \cdot \ldots \cdot \sin\theta_{n-2} \cdot \sin\theta_{n-1},$$

where the θ_ν are subject to the conditions

$$0 \leqslant \theta_\nu \leqslant \pi, \quad 1 \leqslant \nu \leqslant n-2; \quad 0 \leqslant \theta_{n-1} \leqslant 2\pi.$$

With such a choice of parameters, the volume element of the unit sphere is equal to

$$\dot{x} = \sin^{n-2}\theta_1 \cdot \sin^{n-3}\theta_2 \cdot \ldots \cdot \sin\theta_{n-2}\, d\theta_1 \cdot \ldots \cdot d\theta_{n-1}. \qquad (7.2.2)$$

Let us construct an orthogonal system on the unit sphere.

Suppose u is a real vector. We shall consider the fth symmetrized Kronecker power $u^{[f]}$ of the vector u, i.e., the vector of dimension

$$N_f = \frac{1}{f!} n(n+1)\ldots(n+f-1)$$

and components

$$\sqrt{\frac{f!}{j_1!\ldots j_n!}}\, u_1^{j_1}\ldots u_n^{j_n}, \qquad j_1 + \ldots + j_n = f.$$

It can be readily proved that

$$u^{[f]}(v^{[f]})' = (uv')^f.$$

Hence

$$x^{[f]}(x^{[f]})' = 1.$$

Let us now consider the real orthogonal transformation

$$v = uT, \quad TT' = 1. \qquad (7.2.3)$$

The relation

$$v^{[f]} = u^{[f]}T^{[f]}$$

holds.

Let us now decompose the vector space $u^{[f]}$ into a direct sum of irreducible invariant subspaces. It is evident that the vectors

$$(uu')\,u^{[f-2]}$$

form an N_{f-2}-dimensional invariant subspace of the space $u^{[f]}$. Hence one can prove that $T^{[f]}$ can be decomposed into the direct sum

$$T^{[f]} = T_f \dot{+} T_{f-2} \dot{+} \cdots \dot{+} T_{f-2\left[\frac{f}{2}\right]}, \tag{7.2.4}$$

where T_ν is a square matrix of order $N_\nu - N_{\nu-2}$. It is well known that the T_ν are irreducible and pairwise inequivalent (see Murnaghan [1, pp. 328-329]), and we can assume that they are orthogonal. We shall denote by $V_{f-2l}(u)$ the projection of the vector $u^{[f]}$ in the invariant subspace corresponding to the term T_{f-2l} in (7.2.4), and let

$$(uu')^l \varphi^{(i)}_{f-2l}(u), \qquad i = 1, 2, \ldots, N_{f-2l} - N_{f-2l-2},$$

be the components of the vector $V_{f-2l}(u)$ in this invariant subspace.
Applying Schur's lemma as usual, we obtain:

THEOREM 7.2.1. *We have*

$$\int_{\Upsilon} \varphi^{(i)}_\nu(x)\,\varphi^{(j)}_\mu(x)\,\dot{x} = \beta_\nu \cdot \delta_{ij}\delta_{\nu\mu}, \tag{7.2.5}$$

where β_ν does not depend on i.

Let us set

$$\psi^{(i)}_\nu(u) = \beta_\nu^{-\frac{1}{2}}\,\varphi^{(i)}_\nu(u). \tag{7.2.6}$$

These functions form on the sphere an orthonormal system.

7.3. **Projection of a kernel into a subspace.** Let us consider the function

$$\Phi_\nu(u, v) = \sum_{i=1}^{N_\nu - N_{\nu-2}} \psi^{(i)}_\nu(u)\,\psi^{(i)}_\nu(v) = V_\nu(u)\,V'_\nu(v). \tag{7.3.1}$$

It is evident that for any orthogonal matrix T

$$\Phi_\nu(uT, vT) = V_\nu(uT)\,V'_\nu(vT) = V_\nu(u)\,T_\nu T'_\nu V'_\nu(v) = V_\nu(u)\,V'_\nu(v) = \Phi_\nu(u, v),$$

so that $\Phi_\nu(u,v)$ is invariant under orthogonal transformations. Hence, by the corresponding theorem of the theory of invariants (see H. Weyl [2, p. 53]), $\Phi(u,v)$ can be represented as a function of uu', vv' and uv'. Since

$\Phi_\nu(u, v)$ is a homogeneous function of the coordinates of each of the vectors u and v of order ν, we can write

$$\Phi_\nu(u, v) = \sum_{l=0}^{[\nu/2]} c_{l,\nu} (uv')^{\nu-2l} (uu'vv')^l.$$

If $u=x$ and $v=y$ are points on the sphere, then

$$\Phi_\nu(x, y) = \sum_{l=0}^{[\nu/2]} c_{l,\nu} (xy')^{\nu-2l}.$$

We set

$$\Phi_\nu(x, y) = Q_\nu(\xi), \quad \xi = xy'. \tag{7.3.2}$$

It is evident that ξ is equal to the cosine of the angle between the rays directed from the origin to the points x and y.

From (7.3.1) and (7.2.5) we have

$$\int_\Upsilon \int_\Upsilon \Phi_\nu(x, y) \Phi_\mu(x, y) \dot{x} \dot{y} = (N_\nu - N_{\nu-2}) \delta_{\mu\nu}. \tag{7.3.3}$$

Substituting (7.3.2) in (7.3.3), we obtain

$$\int_\Upsilon \int_\Upsilon Q_\nu(xy') Q_\mu(xy') \dot{x} \dot{y} = (N_\nu - N_{\nu-2}) \delta_{\mu\nu}. \tag{7.3.4}$$

For a fixed x, there exists an orthogonal matrix T with determinant $+1$ such that $xT = (1, 0, \cdots, 0)$. Setting $w = yT$, we find that the integral (7.3.4) is equal to $[w = (w_1, \cdots, w_n)]$:

$$\int_\Upsilon \int_\Upsilon Q_\nu(w_1) Q_\mu(w_1) \dot{w} \dot{x} = \omega \int_\Upsilon Q_\nu(w_1) Q_\mu(w_1) \dot{w},$$

where ω is the area of the surface of the n-dimensional unit sphere. We effect a change of variables:

$$w_1 = \xi, \quad w_i = (1 - \xi^2)^{\frac{1}{2}} \xi_i, \qquad 2 \leqslant i \leqslant n.$$

Then

$$\omega \int_{\Upsilon} Q_\nu(w_1) Q_\mu(w_1) \dot{w} = 2\omega \int_{w_1^2+\dots+w_{n-1}^2 \leqslant 1} \dots \int Q_\nu(w_1) Q_\mu(w_1) \frac{dw_1 \dots dw_{n-1}}{\sqrt{1-w_1^2-\dots-w_{n-1}^2}}$$

$$= 2\omega \int_{-1}^{1} Q_\nu(w_1) Q_\mu(w_1)\, dw_1 \cdot \int_{w_2^2+\dots+w_{n-1}^2 \leqslant 1-w_1^2} \dots \int \frac{dw_2 \dots dw_{n-1}}{\sqrt{1-w_1^2-\dots-w_{n-1}^2}}$$

$$= 2\omega \int_{-1}^{1} Q_\nu(\xi) Q_\mu(\xi) (1-\xi^2)^{\frac{n-3}{2}} d\xi \cdot \int_{\xi_2^2+\dots+\xi_{n-1}^2 \leqslant 1} \dots \int \frac{d\xi_2 \dots d\xi_{n-1}}{\sqrt{1-\xi_2^2-\dots-\xi_{n-1}^2}}$$

$$= \omega\omega' \int_{-1}^{1} Q_\nu(\xi) Q_\mu(\xi) (1-\xi^2)^{\frac{n-3}{2}} d\xi, \tag{7.3.5}$$

where ω' is the area of the surface of the $(n-1)$-dimensional unit sphere. From formulas (7.3.4) and (7.3.5) we obtain

$$\int_{-1}^{1} Q_\nu(\xi) Q_\mu(\xi) (1-\xi^2)^{\frac{n-3}{2}} d\xi = \frac{N_\nu - N_{\nu-2}}{\omega\omega'} \delta_{\mu\nu}. \tag{7.3.6}$$

By Theorem 7.1.1

$$Q_\nu(\xi) = c C_\nu^{\frac{n}{2}-1}(\xi). \tag{7.3.7}$$

Let us determine the constant c. From (7.3.6) and (7.1.7) we have

$$\frac{N_\nu - N_{\nu-2}}{\omega\omega'} = c^2 \int_{-1}^{1} \left\{C_\nu^{\frac{n}{2}-1}(\xi)\right\}^2 (1-\xi^2)^{\frac{n-3}{2}} d\xi = c^2 \frac{\pi \cdot 2^{3-n} \cdot \Gamma(\nu+n-2)}{\left[\Gamma\left(\frac{n}{2}-1\right)\right]^2 \nu! \left(\nu+\frac{n}{2}-1\right)}.$$

Hence

$$c^2 = \frac{\Gamma\left(\frac{n}{2}\right) \cdot \Gamma\left(\frac{n-1}{2}\right) \left[\Gamma\left(\frac{n-1}{2}\right)\right]^2 \cdot \nu! \left(\nu+\frac{n}{2}-1\right)}{4\pi^{\frac{n}{2}+\frac{n-1}{2}} \cdot 2^{3-n} \cdot \pi \cdot \Gamma(\nu+n-2)}$$

$$\times \left[\binom{n+\nu-1}{\nu} - \binom{n+\nu-3}{\nu-2}\right] = \frac{1}{4}\pi^{-n} \cdot \left(\nu+\frac{n}{2}-1\right)^2 \cdot \left[\Gamma\left(\frac{n}{2}-1\right)\right]^2,$$

so that

$$c = \frac{1}{2} \cdot \pi^{-\frac{n}{2}} \left(\nu + \frac{n}{2} - 1\right) \Gamma\left(\frac{n}{2} - 1\right). \tag{7.3.8}$$

Thus, finally

$$\Phi_\nu(u, v) = \frac{1}{2} \pi^{-\frac{n}{2}} \left(\nu + \frac{n}{2} - 1\right) \Gamma\left(\frac{n}{2} - 1\right) (uu'vv')^{\frac{\nu}{2}} C_\nu^{\frac{n}{2}-1}\left(\frac{uv'}{\sqrt{uu'vv'}}\right). \tag{7.3.9}$$

7.4. **Orthonormal systems on** $\mathfrak{C}_{\mathrm{IV}}$. The characteristic manifold $\mathfrak{C}_{\mathrm{IV}}$ of the domain $\mathfrak{R}_{\mathrm{IV}}$ consists of points of the form

$$u = e^{i\theta} x, \qquad 0 \leqslant \theta \leqslant \pi, \tag{7.4.1}$$

where x is a real vector which lies on the unit sphere in n-dimensional Euclidean space. For $\dot{u}$, the volume element of $\mathfrak{C}_{\mathrm{IV}}$, we have the expression $\dot{u} = d\theta \cdot \dot{x}$, where $\dot{x}$ is defined by the formula (7.2.2).

Suppose $u^{[f]}$ and $\psi_\nu^{(i)}(u)$ are the same as in §7.2. Then the vector space $u^{[f]}$ can be decomposed as before into a direct sum of invariant subspaces which are defined by the basis vectors

$$\left\{(uu')^l \psi_{f-2l}^{(i)}(u)\right\}, \qquad i = 1, 2, \ldots, N_{f-2l} - N_{f-2l-2}.$$

We readily obtain the theorem:

Theorem 7.4.1. *For $f \neq g$ or $l \neq k$, or $i \neq j$, we have*

$$\int_{\mathfrak{C}_{\mathrm{IV}}} (uu')^l \psi_{f-2l}^{(i)}(u) (\overline{uu'})^k \overline{\psi_{g-2k}^{(j)}(u)}\, \dot{u} = 0. \tag{7.4.2}$$

On the other hand, one can easily verify that

$$\int_{\mathfrak{C}_{\mathrm{IV}}} |uu'|^{2l} |\psi_{f-2l}^{(i)}(u)|^2 \dot{u} = \int_0^\pi d\theta \int_{\Upsilon} |\psi_{f-2l}^{(i)}(x)|^2 \dot{x} = \pi. \tag{7.4.3}$$

Hence the functions

$$\frac{(uu')^l}{\sqrt{\pi}} \psi_{f-2l}^{(i)}(u) \tag{7.4.4}$$

form an orthonormal system on $\mathfrak{C}_{\mathrm{IV}}$.

From this fact, in conjunction with (7.3.9), (7.1.11) and (7.1.12), we obtain

$$H(z, \bar{u}) = \frac{1}{\pi} \sum_{m=0}^{\infty} \sum_{l=0}^{[m/2]} \sum_{i} (zz')^l (\overline{uu'})^l \psi_{m-2l}^{(i)}(z) \overline{\psi_{m-2l}^{(i)}(u)}$$

$$= \frac{1}{\pi} \sum_{m=0}^{\infty} \sum_{l=0}^{[m/2]} (zz'\overline{uu'})^l \Phi_{m-2l}(z, \bar{u})$$

$$= \frac{\Gamma\left(\frac{n}{2}-1\right)}{2\pi^{\frac{n}{2}+1}} \sum_{m=0}^{\infty} \sum_{l=0}^{[m/2]} \left(m-2l+\frac{n}{2}-1\right)(zz'\overline{uu'})^l$$

$$\times (zz'\overline{uu'})^{\frac{m}{2}-l} C_{m-2l}^{\frac{n}{2}-1}\left(\frac{z\bar{u}'}{\sqrt{zz'\overline{uu'}}}\right)$$

$$= \left(\frac{n}{2}-1\right)\frac{\Gamma\left(\frac{n}{2}-1\right)}{2\pi^{\frac{n}{2}+1}} \sum_{m=0}^{\infty} (zz'\overline{uu'})^{\frac{m}{2}} C_m^{\frac{n}{2}}\left(\frac{z\bar{u}'}{\sqrt{zz'\overline{uu'}}}\right)$$

$$= \frac{1}{2}\pi^{-\frac{n}{2}-1}\Gamma\left(\frac{n}{2}\right)\cdot(1-2z\bar{u}'+zz'\overline{uu'})^{-\frac{n}{2}}. \tag{7.4.5}$$

REMARK 1. As $uu' = e^{2i\theta}$, $\bar{u}'e^{2i\theta} = u'$, it follows that

$$1-2z\bar{u}'+zz'\overline{uu'} = \overline{uu'}\,(uu'-2z\bar{u}'uu'+zz')$$

$$= \overline{uu'}\,(uu'-2zu'+zz') = \overline{uu'}\,(u-z)(u-z)',$$

so that Cauchy's formula can be written in the form

$$f(z) = \frac{\Gamma\left(\frac{n}{2}\right)}{2\pi^{\frac{n}{2}+1}} \int_0^{\pi}\int_{\gamma} \frac{f(e^{i\theta}x)\, e^{in\theta}}{[(e^{i\theta}x-z)(e^{i\theta}x-z)']^{\frac{n}{2}}} d\theta \cdot \dot{x}$$

$$= \frac{\Gamma\left(\frac{n}{2}\right)}{2\pi^{\frac{n}{2}+1}} \int_0^{\pi}\int_{\gamma} \frac{f(e^{i\theta}x)}{[(x-e^{-i\theta}z)(x-e^{-i\theta}z)']^{\frac{n}{2}}} d\theta \cdot \dot{x}. \tag{7.4.6}$$

7.5. **A complete orthonormal system in $\mathfrak{R}_{\text{IV}}$.** We already know that the functions

$$(zz')^l \psi_{f-2l}^{(i)}(z)$$

form in $\Re_{IV}$ a complete orthonormal system. We shall now proceed to the determination of the constant

$$\tau_{l,f-2l}=\int_{\Re_{IV}} |zz'|^{2l}\,|\psi^{(i)}_{f-2l}(z)|^2\,\dot{z},$$

or, what amounts to the same,

$$\tau_{l,f-2l}=\frac{1}{N_{f-2l}-N_{f-2l-2}}\int_{\Re_{IV}} |zz'|^{2l}\,\Phi_{f-2l}(z,\bar{z})\,\dot{z}. \qquad (7.5.1)$$

If we start with the calculation of the integral

$$\int_{\Re_{IV}} (z\bar{z}')^s\,|z\bar{z}'|^{2t}\,\dot{z},$$

the problem turns out to be not at all simple. Therefore we shall prefer a less direct method. From the definition of the Bergman kernel and from its expression for $\Re_{IV}$ it follows that

$$\frac{1}{V(\Re_{IV})}(1-2z\bar{z}'+|zz'|^2)^{-n}=\sum_{m=0}^{\infty}\sum_{l=0}^{[m/2]}\frac{1}{\tau_{l,m-2l}}|zz'|^{2l}\,\Phi_{m-2l}(z,\bar{z})$$

$$=\sum_{m=0}^{\infty}\sum_{l=0}^{[m/2]}\frac{|zz'|^m\left(m-2l+\frac{n}{2}-1\right)\Gamma\left(\frac{n}{2}-1\right)}{\tau_{l,m-2l}\cdot 2\pi^{\frac{n}{2}}}C^{\frac{n}{2}-1}_{m-2l}(\xi)$$

$$=\frac{1}{2}\pi^{-\frac{n}{2}}\Gamma\left(\frac{n}{2}-1\right)\sum_{m=0}^{\infty}|zz'|^m\sum_{l=0}^{[m/2]}\frac{m-2l+\frac{n}{2}-1}{\tau_{l,m-2l}}C^{\frac{n}{2}-1}_{m-2l}(\xi),$$

where $\xi=z\bar{z}'/|zz'|$.

On the other hand, by (7.1.2) we have

$$\sum_{m=0}^{\infty}w^m C^{\lambda}_m(\xi)=(1-2w\xi+w^2)^{-\lambda},$$

whence

$$\frac{1}{V(\Re_{IV})}(1-2z\bar{z}'+|zz'|^2)^{-n}=\frac{1}{V(\Re_{IV})}\sum_{m=0}^{\infty}|zz'|^m\,C^n_m(\xi).$$

By comparing the coefficients, we obtain

$$\frac{1}{V(\Re_{IV})} C_m^n(\xi) = \frac{1}{2}\pi^{-\frac{n}{2}}\Gamma\left(\frac{n}{2}-1\right)\sum_{l=0}^{[m/2]}\frac{m-2l+\frac{n}{2}-1}{\tau_{l,\,m-2l}} C_{m-2l}^{\frac{n}{2}-1}(\xi).$$

By (7.1.11)

$$C_m^{\nu}(\xi) = \frac{\Gamma(\lambda)}{\Gamma(\nu)}\sum_{k=0}^{[m/2]} c_k C_{m-2k}^{\lambda}(\xi),$$

where

$$c_k = \frac{(m-2k+\lambda)\,\Gamma(k+\nu-\lambda)\,\Gamma(m+\nu-k)}{k!\,\Gamma(\nu-\lambda)\,\Gamma(m-k+\lambda+1)}.$$

Hence

$$\frac{\Gamma\left(\frac{n}{2}-1\right)\left(m-2l+\frac{n}{2}-1\right)\Gamma\left(l+\frac{n}{2}+1\right)\Gamma(m+n-l)}{V(\Re_{IV})\,\Gamma(n)\,l!\,\Gamma\left(\frac{n}{2}+1\right)\Gamma\left(m-l+\frac{n}{2}\right)} = \frac{\Gamma\left(\frac{n}{2}-1\right)\left(m-2l+\frac{n}{2}-1\right)}{2\pi^{\frac{n}{2}}\tau_{l,\,m-2l}},$$

and we obtain

$$\tau_{l,\,m-2l} = \frac{l!\,\Gamma(n)\,\Gamma\left(\frac{n}{2}+1\right)\Gamma\left(m+\frac{n}{2}-l\right)}{2\pi^{\frac{n}{2}}\,\Gamma\left(l+\frac{n}{2}+1\right)\Gamma(m+n-l)} V(\Re_{IV}). \tag{7.5.2}$$

In conclusion, let us write

$$\int_{\Re_{IV}} |zz'|^{2l}\,\Phi_{f-2l}(z,\bar{z})\,\dot{z}$$

$$= (N_{f-2l}-N_{f-2l-2})\frac{l!\,\Gamma(n)\,\Gamma\left(\frac{n}{2}+1\right)\Gamma\left(f+\frac{n}{2}-l\right)}{2\pi^{\frac{n}{2}}\,\Gamma\left(l+\frac{n}{2}+1\right)\Gamma(f+n-l)} V(\Re_{IV}). \tag{7.5.3}$$

Since

$$N_{f-2l} = \frac{(n+f-2l-1)!}{(f-2l)!\,(n-1)!}, \qquad V(\Re_{IV}) = \frac{\pi^n}{2^{n-1}\,n!},$$

and

$$\Phi_{f-2l}(z,\bar{z}) = \frac{1}{2}\pi^{-\frac{n}{2}}\left(f-2l+\frac{n}{2}-1\right)\Gamma\left(\frac{n}{2}-1\right)|zz'|^{f-2l}C_{f-2l}^{\frac{n}{2}-1}\left(\frac{z\bar{z}'}{|zz'|}\right),$$

it follows that formula (7.5.3) is equivalent to the following formula

$$\int_{\mathfrak{R}_{IV}} | zz' |^f C_{f-2l}^{\frac{n}{2}-1}\left(\frac{z\bar{z}'}{| zz' |}\right) \cdot \dot{z}$$

$$= \left(\frac{\pi}{2}\right)^n \frac{(n+f-2l-3)!}{(f-2l)!\,(n-3)!} \cdot \frac{l!\,\Gamma\left(f+\frac{n}{2}-l\right)}{\Gamma\left(l+\frac{n}{2}+1\right)\Gamma(f+n-l)}. \tag{7.5.4}$$

Below we shall obtain a direct proof of formula (7.5.4).

7.6. **Reduction of a multiple integral to a simple one.** As we know, the domain $\mathfrak{R}_{IV}$ can be defined by the inequality

$$xx' + yy' + 2\sqrt{xx'yy' - (xy')^2} < 1,$$

where we set $z = x + iy$ (see §2.5).

Let us consider the integral

$$I = \int_{\mathfrak{R}_{IV}} | zz' |^f F\left(\frac{z\bar{z}'}{| zz' |}\right) \dot{z}. \tag{7.6.1}$$

With fixed x, we effect a change of variables

$$y = \sqrt{xx'}\, u.$$

Then, evidently, $\dot{y} = (xx')^{n/2}\,\dot{u}$. The domain of integration will now be determed by the inequality

$$xx'\left(1 + uu' + 2\sqrt{uu' - \left(\frac{xu'}{\sqrt{xx'}}\right)^2}\right) < 1.$$

We shall take an orthogonal matrix Γ with determinant equal to 1 such that

$$x\Gamma = (\sqrt{xx'},\ 0,\ \ldots,\ 0),$$

and we shall effect the change of variables

$$u\Gamma = (\xi,\ v),$$

where v is an $(n-1)$-dimensional vector, $v = (v_2, \cdots, v_n)$. Since

$$z\bar{z}' = xx' + yy' = xx'(1 + \xi^2 + vv')$$

and

$$| zz' |^2 = (xx' - yy')^2 + 4(xy')^2 = (xx')^2\{(1 - \xi^2 - vv')^2 + 4\xi^2\},$$

it follows that

$$I = \int\limits_{xx'(1+\xi^2+vv'+2\sqrt{vv'})<1} (xx')^{f+\frac{n}{2}} \{(1-\xi^2-vv')^2+4\xi^2\}^{\frac{f}{2}}$$

$$\times F\left(\frac{1+\xi^2+vv'}{[(1-\xi^2-vv')^2+4\xi^2]^{1/2}}\right) d\xi \cdot \dot{v} \cdot \dot{x}.$$

Further, by means of the relation

$$\int\limits_{xx'\leqslant A^2} (xx')^{f+\frac{n}{2}} \dot{x} = \frac{2\pi^{\frac{n}{2}}}{\Gamma\left(\frac{n}{2}\right)} \int_0^A r^{n-1} \cdot r^{2f+n}\, dr = \frac{\pi^{\frac{n}{2}}}{\Gamma\left(\frac{n}{2}\right)} \cdot \frac{A^{2(f+n)}}{f+n},$$

we obtain

$$I = \frac{\pi^{\frac{n}{2}}}{(f+n)\Gamma\left(\frac{n}{2}\right)} \int_{-\infty}^{\infty} \cdots \int \left(1+\xi^2+vv'+2\sqrt{vv'}\right)^{-(f+n)}$$

$$\times [(1-\xi^2-vv')^2+4\xi^2]^{\frac{f}{2}} F\left(\frac{1+\xi^2+vv'}{[(1-\xi^2-vv')^2+4\xi^2]^{1/2}}\right) d\xi \cdot \dot{v}$$

$$= \frac{\pi^{\frac{n}{2}} \cdot 2^n}{(f+n)\Gamma\left(\frac{n}{2}\right)} \int\limits_{v_i\geqslant 0,\ \xi\geqslant 0} \cdots \int (*)\, d\xi \cdot \dot{v}. \tag{7.6.2}$$

We shall set $vv' = \eta^2$, $\eta > 0$, $v_2 = \sqrt{\eta^2 - v_3^2 - \cdots - v_n^2}$. Then

$$\dot{v} = \frac{\eta\, d\eta\, dv_3 \ldots dv_n}{\sqrt{\eta^2 - v_3^2 - \ldots - v_n^2}},$$

and we obtain

$$I = \frac{2^n \pi^{\frac{n}{2}}}{(f+n)\Gamma\left(\frac{n}{2}\right)} \int\limits_{\substack{v_i\geqslant 0,\ i=3,\ldots,n\\ \xi\geqslant 0,\ \eta\geqslant 0}} \cdots \int (1+\xi^2+\eta^2+2\eta)^{-(f+n)} \{(1-\xi^2-\eta^2)+4\xi^2\}^{\frac{f}{2}}$$

$$\times F\left(\frac{1+\xi^2+\eta^2}{\sqrt{(1-\xi^2-\eta^2)^2+4\xi^2}}\right) \frac{\eta\, d\eta\, d\xi\, dv_3 \ldots dv_n}{\sqrt{\eta^2 - v_3^2 - \ldots - v_n^2}}.$$

Since

$$\int \cdots \int\limits_{\substack{v_3^2 + \cdots + v_n^2 \leqslant \eta^2 \\ v_3 \geqslant 0, \ldots, v_n \geqslant 0}} \frac{dv_3 \ldots dv_n}{\sqrt{\eta^2 - v_3^2 - \cdots - v_n^2}}$$

$$= \frac{\eta^{n-3}}{2^{n-2}} \int \cdots \int\limits_{v_3^2 + \cdots + v_n^2 \leqslant 1} \frac{dv_3 \ldots dv_n}{\sqrt{1 - v_3^2 - \cdots - v_n^2}} = \frac{\eta^{n-3}}{2^{n-2}} \cdot \frac{\pi^{\frac{n-1}{2}}}{\Gamma\left(\frac{n-1}{2}\right)},$$

we obtain

$$I = \frac{2^n \pi^{\frac{n}{2}}}{(f+n)\Gamma\left(\frac{n}{2}\right)} \cdot \frac{\pi^{\frac{n-1}{2}}}{2^{n-2}\Gamma\left(\frac{n-1}{2}\right)} \int_0^\infty \int_0^\infty (1 + \xi^2 + \eta^2 + 2\eta)^{-(f+n)}$$

$$\times \{(1 - \xi^2 - \eta^2)^2 + 4\xi^2\}^{\frac{f}{2}} F\left(\frac{1 + \xi^2 + \eta^2}{\sqrt{(1 - \xi^2 - \eta^2)^2 + 4\xi^2}}\right) \eta^{n-2}\, d\xi\, d\eta$$

$$= \frac{2^n \pi^{n-1}}{(f+n)\Gamma(n-1)} \int_0^\infty \int_0^\infty (1 + \xi^2 + \eta^2 + 2\eta)^{-(f+n)}$$

$$\times \{(1 - \xi^2 - \eta^2)^2 + 4\xi^2\}^{\frac{f}{2}} F\left(\frac{1 + \xi^2 + \eta^2}{\sqrt{(1 - \xi^2 - \eta^2)^2 + 4\xi^2}}\right) \eta^{n-2}\, d\xi\, d\eta$$

[we used the formula $\Gamma(x)\Gamma(x+\frac{1}{2}) = \pi^{1/2}\Gamma(2x)2^{1-2x}$]. Let us note that

$$(1 - \xi^2 - \eta^2)^2 + 4\xi^2 = (1 + \xi^2 + \eta^2)^2 - 4\eta^2,$$

and effect the change of variables $\tau = (1+\xi^2+\eta^2)/2\eta$. Then

$$\frac{1+\eta^2}{2\eta} \leqslant \tau < \infty, \qquad d\tau = \frac{\xi}{\eta}\, d\xi, \qquad d\xi = \frac{\eta\, d\tau}{\sqrt{2\eta\tau - 1 - \eta^2}},$$

and we obtain

$$I = \frac{2^n \pi^{n-1}}{(f+n)\Gamma(n-1)} \int_0^\infty \frac{d\eta}{\eta} \int_{\frac{1+\eta^2}{2\eta}}^\infty 2^{-n}(1+\tau)^{-(f+n)}$$

$$\times (\tau^2 - 1)^{\frac{f}{2}} F\left(\frac{\tau}{\sqrt{\tau^2 - 1}}\right) \frac{d\tau}{\sqrt{2\eta\tau - 1 - \eta^2}}.$$

But

$$\int\limits_{2\eta\tau-1-\eta^2\geqslant 0}\frac{d\eta}{\eta\sqrt{2\eta\tau-1-\eta^2}}=\pi,$$

and we finally obtain

$$I=\frac{\pi^n}{(f+n)\Gamma(n-1)}\int\limits_1^\infty(\tau+1)^{-(f+n)}(\tau^2-1)^{\frac{f}{2}}F\left(\frac{\tau}{\sqrt{\tau^2-1}}\right)d\tau. \tag{7.6.3}$$

7.7. **Another form of the expression** (7.6.3). We shall set

$$t=\frac{\tau}{\sqrt{\tau^2-1}},\qquad d\tau=-(t^2-1)^{-\frac{3}{2}}dt.$$

Then (7.6.3) assumes the form

$$I=\frac{\pi^n}{(f+n)\Gamma(n-1)}\int\limits_1^\infty\left(t+\sqrt{t^2-1}\right)^{-(f+n)}(t^2-1)^{\frac{n-3}{2}}F(t)\,dt. \tag{7.7.1}$$

In order to prove now (7.5.4), it suffices to establish that

$$\int\limits_1^\infty\left(t+\sqrt{t^2-1}\right)^{-(2l+m+n)}(t^2-1)^{\frac{n-3}{2}}C_m^{\frac{n}{2}-1}(t)\,dt$$

$$=\frac{(2l+m+n)(n-2)}{2^n}\cdot\frac{l!\,(n+m-3)!\,\Gamma\left(m+\frac{n}{2}+l\right)}{m!\,\Gamma\left(l+\frac{n}{2}+1\right)\Gamma(m+n+l)}. \tag{7.7.2}$$

By Rodrigues' formula we have

$$(1-t^2)^{\frac{n-3}{2}}C_m^{\frac{n}{2}-1}(t)$$

$$=\frac{(-2)^m}{m!}\cdot\frac{\Gamma\left(m+\frac{n}{2}-1\right)\Gamma(m+n-2)}{\Gamma\left(\frac{n}{2}-1\right)\Gamma(2m+n-2)}\cdot\left(\frac{d}{dt}\right)^m(1-t^2)^{m+\frac{n-3}{2}}.$$

Therefore (7.7.2) reduces to the form

$$\int\limits_1^\infty\left(t+\sqrt{t^2-1}\right)^{-(2l+m+n)}\left(\frac{d}{dt}\right)^m(t^2-1)^{m+\frac{n-3}{2}}dt$$

$$=\frac{\Gamma\left(\frac{n}{2}\right)}{2^{m+n-1}}\cdot\frac{(2l+m+n)\Gamma\left(m+\frac{n}{2}+l\right)\Gamma(2m+n-2)\,l!}{\Gamma\left(l+\frac{n}{2}+1\right)\Gamma(m+n+l)\Gamma\left(m+\frac{n}{2}-1\right)}. \tag{7.7.3}$$

This formula does not contain special functions, but its direct proof presents considerable difficulties.

Let us now effect in (7.7.1) the change of variables $t=\operatorname{ch} x$. Thereupon we obtain

$$I=\frac{\pi^n}{(f+n)\Gamma(n-1)}\int_0^\infty e^{-x(f+n)}(\operatorname{sh} x)^{n-2}F(\operatorname{ch} x)\,dx. \qquad (7.7.4)$$

Expression (7.5.4) is equivalent to the following formula:

$$\int_0^\infty e^{-x(m+n+2l)}(\operatorname{sh} x)^{n-2}C_m^{\frac{n}{2}-1}(\operatorname{ch} x)\,dx$$

$$=\frac{(2l+m+n)(n-2)}{2^n}\cdot\frac{l!(m+n-3)!\,\Gamma\left(m+\frac{n}{2}+l\right)}{m!\,\Gamma\left(l+\frac{n}{2}+1\right)\Gamma(m-n+l)}. \qquad (7.7.5)$$

7.8. **Proof of formula** (7.7.5). Let us denote

$$a^{(q)}=a(a+1)\ldots(a+q-1).$$

Evidently

$$\frac{\Gamma(a+q)}{\Gamma(a)}=a^{(q)},\quad \frac{\Gamma(a)}{\Gamma(a-q)}=(-1)^q(1-a)^{(q)},$$

$$2^{-2q}\frac{\Gamma(a+2q)}{\Gamma(a)}=\left(\frac{a}{2}\right)^{(q)}\left(\frac{a+1}{2}\right)^{(q)}$$

The generalized hypergeometric series is defined (see Bailey [1])[20] by the formula

$${}_{p+1}F_p\begin{pmatrix}\alpha_1,\ldots,\alpha_{p+1};\ z\\ \beta_1,\ldots,\beta_p\end{pmatrix}=\sum_{q=0}^\infty\frac{(\alpha_1)^{(q)}\ldots(\alpha_{p+1})^{(q)}}{(\beta_1)^{(q)}\ldots(\beta_p)^{(q)}}\cdot\frac{z^q}{q!}. \qquad (7.8.1)$$

LEMMA 1. *For* $s>l$

$$I_1=\int_0^\infty e^{-xs}(\operatorname{sh} x)^l\,dx=\frac{\Gamma(l+1)}{2^{l+1}}\,\frac{\Gamma\left(\frac{s-l}{2}\right)}{\Gamma\left(\frac{s+l}{2}+1\right)}.$$

PROOF. Effecting the change of variables $y=e^{-2x}$, we obtain

[20] Translator's note. See also Erdelyi, Magnus et al., *Higher transcendental functions*, Vol. I, 1953.

$$I_1=\frac{1}{2^l}\int_0^\infty e^{-(s-l)x}(1-e^{-2x})^l\,dx=\frac{1}{2^{l+1}}\int_0^1 y^{\frac{s-l}{2}-1}(1-y)^l\,dy$$

$$=\frac{1}{2^{l+1}}\,\frac{\Gamma(l+1)\,\Gamma\left(\frac{s-l}{2}\right)}{\Gamma\left(\frac{s+l}{2}+1\right)}.$$

LEMMA 2. *For* $s>l+1$

$$I_2=\int_0^\infty e^{-xs}\operatorname{ch} x\,(\operatorname{sh} x)^l\,dx=s\,\frac{\Gamma(l+1)}{2^{l+2}}\,\frac{\Gamma\left(\frac{s-l}{2}-\frac{1}{2}\right)}{\Gamma\left(\frac{s+l}{2}+\frac{3}{2}\right)}.$$

PROOF. By Lemma 1, we have

$$I_2=\frac{1}{2}\int_0^\infty e^{-x(s-1)}(\operatorname{sh} x)^l\,dx+\frac{1}{2}\int_0^\infty e^{-x(s+1)}(\operatorname{sh} x)^l\,dx$$

$$=\frac{\Gamma(l+1)}{2^{l+2}}\left\{\frac{\Gamma\left(\frac{s-l-1}{2}\right)}{\Gamma\left(\frac{s+l-1}{2}+1\right)}+\frac{\Gamma\left(\frac{s-l+1}{2}\right)}{\Gamma\left(\frac{s+l+1}{2}+1\right)}\right\}$$

$$=\frac{\Gamma(l+1)}{2^{l+2}}\,\frac{\Gamma\left(\frac{s-l-1}{2}\right)}{\Gamma\left(\frac{s+l+3}{2}\right)}\left(\frac{s+l+1}{2}+\frac{s-l-1}{2}\right).$$

LEMMA 3. *If one of the numbers* $\alpha_1,\cdots,\alpha_{p+1}$ *is a negative integer, then*

$$\int_0^\infty e^{-xs}(\operatorname{sh} x)^{2\lambda}\,{}_{p+1}F_p\left(\begin{matrix}\alpha_1,\ \ldots,\ \alpha_{p+1};\ -\operatorname{sh}^2 x\\ \beta_1,\ \ldots,\ \beta_p\end{matrix}\right)dx$$

$$=\frac{\Gamma(2\lambda+1)}{2^{2\lambda+1}}\,\frac{\Gamma\left(\frac{s}{2}-\lambda\right)}{\Gamma\left(\frac{s}{2}+\lambda+1\right)}\,{}_{p+3}F_{p+2}\left(\begin{matrix}\alpha_1,\ \ldots,\ \alpha_{p+1},\ \lambda+\frac{1}{2},\ \lambda+1;\ 1\\ \beta_1,\ \ldots,\ \beta_p,\ \frac{s}{2}+\lambda+1,\ \lambda+1-\frac{s}{2}\end{matrix}\right).$$

PROOF. We substitute in the left-hand side of the formula the expression (7.8.1) and integrate by parts. Since the general term of the series so obtained is equal to

$$(-1)^q\,\frac{(\alpha_1)^{(q)}\ \ldots\ (\alpha_{p+1})^{(q)}}{q!\,(\beta_1)^{(q)}\ \ldots\ (\beta_p)^{(q)}}\int_0^\infty(\operatorname{sh} x)^{2\lambda+2q}\,e^{-xs}\,dx$$

$$=(-1)^q \cdot \frac{(\alpha_1)^{(q)} \ldots (\alpha_{p+1})^{(q)}}{q!\,(\beta_1)^{(q)} \ldots (\beta_p)^{(q)}} \cdot \frac{\Gamma(2\lambda+2q+1)}{2^{2\lambda+2q+1}} \cdot \frac{\Gamma\left(\frac{s}{2}-\lambda-q\right)}{\Gamma\left(\frac{s}{2}+\lambda+q+1\right)}$$

$$=\frac{(\alpha_1)^{(q)} \ldots (\alpha_{p+1})^{(q)}}{q!\,(\beta_1)^{(q)} \ldots (\beta_p)^{(q)}} \cdot \frac{\left(\lambda+\frac{1}{2}\right)^{(q)}(\lambda+1)^{(q)}}{\left(\frac{s}{2}+\lambda+1\right)^{(q)}\left(\lambda-\frac{s}{2}+1\right)^{(q)}} \cdot \frac{\Gamma(2\lambda+1)}{2^{2\lambda+1}} \frac{\Gamma\left(\frac{s}{2}-\lambda\right)}{\Gamma\left(\frac{s}{2}+\lambda+1\right)},$$

we obtain the assertion of the lemma.

It is well known (see Szegö [1, p. 84]) that

$$C_{2\nu}^{\lambda}(x)=\frac{\Gamma(2\nu+2\lambda)}{(2\nu)!\,\Gamma(2\lambda)} \cdot {}_2F_1\left(\begin{matrix}-\nu, & \nu+\lambda; & 1-x^2\\ & \lambda+\frac{1}{2} & \end{matrix}\right).$$

By Lemma 3 we have

$$\int_0^\infty (\operatorname{sh} x)^{2\lambda} C_{2\nu}^{\lambda}(\operatorname{ch} x)\, e^{-xs}\, dx$$

$$=\frac{\Gamma(2\nu+2\lambda)}{(2\nu)!\,\Gamma(2\lambda)} \int_0^\infty (\operatorname{sh} x)^{2\lambda}\, {}_2F_1\left(\begin{matrix}-\nu, & \nu+\lambda; & -\operatorname{sh}^2 x\\ & \lambda+\frac{1}{2} & \end{matrix}\right) e^{-xs}\, dx$$

$$=\frac{\Gamma(2\nu+2\lambda)\,\Gamma(2\lambda+1)}{(2\nu)!\,\Gamma(2\lambda)\quad 2^{2\lambda+1}} \cdot \frac{\Gamma\left(\frac{s}{2}-\lambda\right)}{\Gamma\left(\frac{s}{2}+\lambda+1\right)}\, {}_4F_3\left(\begin{matrix}-\nu, \nu+\lambda, \lambda+\frac{1}{2}, \lambda+1; 1\\ \lambda+\frac{1}{2}, \frac{s}{2}+\lambda+1, \lambda-\frac{s}{2}+1\end{matrix}\right)$$

$$=\frac{\Gamma(2\nu+2\lambda)}{(2\nu)!\,\Gamma(2\lambda)} \frac{\Gamma(2\lambda+1)}{2^{2\lambda+1}} \frac{\Gamma\left(\frac{s}{2}-\lambda\right)}{\Gamma\left(\frac{s}{2}+\lambda+1\right)}\, {}_3F_2\left(\begin{matrix}-\nu, & \nu+\lambda, & \lambda+1; \ 1\\ \frac{s}{2}+\lambda+1, & \lambda-\frac{s}{2}+1 & \end{matrix}\right),$$

where ${}_3F_2$ satisfies Saalschütz's condition (see Bailey [1, p. 9]), i.e., $a+b+c+1=d+e$, and at least one of the a, b, c is a negative integer. Then (Saalschütz's theorem)

$${}_3F_2\left(\begin{matrix}a, & b, & c; & 1\\ & d, & e & \end{matrix}\right)=\frac{\Gamma(d)\,\Gamma(a-e+1)\,\Gamma(b-e+1)\,\Gamma(c-e+1)}{\Gamma(1-e)\,\Gamma(d-a)\,\Gamma(d-b)\,\Gamma(d-c)}.$$ [21]

Therefore

---

[21] Translator's note. See also note on p. 155.

$$\int_0^\infty (\operatorname{sh} x)^{2\lambda} C_{2\nu}^\lambda(\operatorname{ch} x)\,dx = \frac{\Gamma(2\nu+2\lambda)}{(2\nu)!\,\Gamma(2\lambda)} \cdot \frac{\Gamma(2\lambda+1)}{2^{2\lambda+1}}$$

$$\times \frac{\Gamma\left(\frac{s}{2}-\lambda\right)}{\Gamma\left(\frac{s}{2}+\lambda+1\right)} \cdot \frac{\Gamma\left(\frac{s}{2}+\lambda+1\right)}{\Gamma\left(\frac{s}{2}-\lambda\right)} \cdot \frac{\Gamma\left(\frac{s}{2}-\lambda-\nu\right)\Gamma\left(\frac{s}{2}+\nu\right)\Gamma\left(\frac{s}{2}+1\right)}{\Gamma\left(\frac{s}{2}+\lambda+\nu+1\right)\Gamma\left(\frac{s}{2}-\nu+1\right)\Gamma\left(\frac{s}{2}\right)},$$

and formula (7.7.5) for even m is proved.

Lemma 4. *If at least one of the* $\alpha_1, \cdots, \alpha_{p+1}$ *is a negative integer, then*

$$\int_0^\infty (\operatorname{sh} x)^{2\lambda}\, {}_{p+1}F_p\begin{pmatrix}\alpha_1, \ldots, \alpha_{p+1}; & -\operatorname{sh}^2 x\\ \beta_1, \ldots, \beta_p & \end{pmatrix} e^{-sx} \operatorname{ch} x\,dx$$

$$= s \cdot \frac{\Gamma(2\lambda+1)}{2^{2\lambda+2}} \cdot \frac{\Gamma\left(\frac{s-1}{2}-\lambda\right)}{\Gamma\left(\frac{s+3}{2}+\lambda\right)}$$

$$\times {}_{p+3}F_{p+2}\begin{pmatrix}\alpha_1, \ldots, \alpha_{p+1}, \lambda+\frac{1}{2}, \lambda+1; 1\\ \beta_1, \ldots, \beta_p, \lambda-\frac{s}{2}+\frac{3}{2}, \lambda+\frac{s}{2}+\frac{3}{2}\end{pmatrix}.$$

Proof. Substituting in the left-hand side the formula (7.8.1), and integrating by parts, we obtain that the general term is equal to

$$(-1)^q \frac{(\alpha_1)^{(q)} \ldots (\alpha_{p+1})^{(q)}}{q!\,(\beta_1)^{(q)} \ldots (\beta_p)^{(q)}} \int_0^\infty (\operatorname{sh} x)^{2\lambda+2q} \operatorname{ch} x \cdot e^{-xs}\,dx$$

$$= (-1)^q \cdot s \cdot \frac{(\alpha_1)^{(q)} \ldots (\alpha_{p+1})^{(q)}}{q!\,(\beta_1)^{(q)} \ldots (\beta_p)^{(q)}} \cdot \frac{\Gamma(2\lambda+2q+1)}{2^{2\lambda+2q+2}} \cdot \frac{\Gamma\left(\frac{s-1}{2}-\lambda-q\right)}{\Gamma\left(\frac{s+3}{2}+\lambda+q\right)}$$

$$= s\, \frac{(\alpha_1)^{(q)} \ldots (\alpha_{p+1})^{(q)}}{q!\,(\beta_1)^{(q)} \ldots (\beta_p)^{q}} \cdot \frac{\left(\lambda+\frac{1}{2}\right)^{(q)} (\lambda+1)^{(q)}}{\left(\lambda-\frac{s-3}{2}\right)^{(q)} \left(\lambda+\frac{s+3}{2}\right)^{(q)}} \cdot \frac{\Gamma(2\lambda+1)}{2^{2\lambda+2}} \cdot \frac{\Gamma\left(\frac{s-1}{2}-\lambda\right)}{\Gamma\left(\frac{s+3}{2}+\lambda\right)},$$

which proves the lemma.

It is well known that

$$C_{2\nu+1}^\lambda(x) = \frac{\Gamma(2\nu+2\lambda+1)}{(2\nu+1)!\,\Gamma(2\lambda)} \cdot x \cdot {}_2F_1\begin{pmatrix}-\nu, & \nu+\lambda+1, & 1-x^2\\ & \lambda+\frac{1}{2} & \end{pmatrix}.$$

Hence by Lemma 4

$$I = \int_0^\infty (\operatorname{sh} x)^{2\lambda} C_{2\nu+1}^{\lambda}(\operatorname{ch} x)\, e^{-xs}\, dx = \frac{\Gamma(2\nu+2\lambda+1)}{(2\nu+1)!\,\Gamma(2\lambda)}$$

$$\times s \cdot \frac{\Gamma(2\lambda+1)}{2^{2\lambda+2}} \cdot \frac{\Gamma\left(\frac{s-1}{2}-\lambda\right)}{\Gamma\left(\frac{s+3}{2}+\lambda\right)} {}_4F_3\left(\begin{matrix} -\nu,\ \nu+\lambda+1,\ \lambda+\frac{1}{2},\ \lambda+1;\ 1 \\ \lambda+\frac{1}{2},\ \lambda-\frac{s-3}{2},\ \lambda+\frac{s+3}{2} \end{matrix}\right)$$

$$= \frac{\Gamma(2\nu+2\lambda+1)\cdot s\cdot\lambda\cdot\Gamma\left(\frac{s-1}{2}-\lambda\right)}{(2\nu+1)!\,2^{2\lambda+1}\cdot\Gamma\left(\frac{s+3}{2}+\lambda\right)} {}_3F_2\left(\begin{matrix} -\nu,\ \nu+\lambda+1,\ \lambda+1;\ 1 \\ \frac{s+3}{2}+\lambda,\ \lambda-\frac{s-3}{2} \end{matrix}\right).$$

Using Saalschütz's theorem, we obtain

$$I = \frac{s\lambda}{2^{2\lambda+1}} \cdot \frac{\Gamma(2\nu+2\lambda+1)}{(2\nu+1)!} \cdot \frac{\Gamma\left(\frac{s-1}{2}-\lambda\right)}{\Gamma\left(\frac{s+3}{2}+\lambda\right)} \cdot \frac{\Gamma\left(\frac{s+3}{2}+\lambda\right)}{\Gamma\left(\frac{s-1}{2}-\lambda\right)}$$

$$\times \frac{\Gamma\left(\frac{s-1}{2}-\lambda-\nu\right)\Gamma\left(\frac{s+1}{2}+\nu\right)\Gamma\left(\frac{s+1}{2}\right)}{\Gamma\left(\frac{s+3}{2}+\lambda+\nu\right)\Gamma\left(\frac{s+1}{2}-\nu\right)\Gamma\left(\frac{s+1}{2}\right)},$$

and the proof of (7.7.5) is complete.

Bibliography

W. N. Bailey

[1] *Generalized hypergeometric series,* Cambridge Univ. Press, London, 1935.

H. Behnke, P. Thullen

[1] *Theorie der Funktionen mehrerer komplexer Veränderlichen,* Springer, Berlin, 1934.

S. Bergman

[1] *Sur les fonctions orthogonales de plusieurs variables complexes avec les applications à la théorie des fonctions analytiques,* Mémor. Sci. Math. No. 106, Gauthier-Villars, Paris, 1947.

[2] *Kernel functions and extended classes in the theory of functions of complex variables,* Colloque sur les fonctions de plusieurs variables (Bruxelles, 1953), pp. 135-157, Masson, Paris, 1953.

S. Bochner

[1] *Group invariance of Cauchy's formula in several variables,* Ann. of Math. (2) **45** (1944), 686-707.

[2] *A theorem on analytic continuation of functions in several variables,* Ann. of Math. (2) **39** (1938), 14-19.

[3] *Boundary values of analytic functions in several variables and of almost periodic functions,* Ann. of Math. (2) **45** (1944), 708-722.

S. Bochner, W. T. Martin

[1] *Several complex variables,* Princeton Math. Series Vol. 10, Princeton Univ. Press, Princeton, N. J., 1948; Russian transl., IL, Moscow, 1951.

A. Borel

[1] *Les fonctions automorphes de plusieurs variables complexes,* Bull. Soc. Math. France **80**(1952), 167-182.

É. Cartan

[1] *Sur les domaines bornés homogènes de l'espace de n variables complexes,* Abh. Math. Sem. Univ. Hamburg **11**(1935), 116-162.

H. Cartan

[1] *Les fonctions de deux variables complexes et le problème de la représentation analytique,* J. Math. Pures Appl. (9) **10** (1931), 1-114.

A. Erdélyi, W. Magnus, F. Oberhettinger, F. Tricomi

[1] *Higher transcendental functions.* Vol. I, McGraw-Hill, New York, 1953.

B. A. Fuks

[1] *Theory of analytic functions of several complex variables,* OGIZ, Moscow, 1948 (Russian); English transl., Amer. Math. Soc., Providence, R. I., 1963.

L. K. Hua

[1] *On the theory of automorphic functions of a matrix variable.* I. *Geometrical basis,* Amer. J. Math. **66**(1944), 470-488.

[2] *On the theory of automorphic functions of a matrix variable.* II. *The classification of hypercircles under the symplectic group,* Amer. J. Math. **66**(1944), 531-563.

[3] *On the theory of Fuchsian functions of several variables,* Ann. of Math. (2) **47** (1946), 167-191.

[4] *On the theory of functions of several complex variables.* I. *A complete orthonormal system in the hyperbolic space of matrices,* J. Chinese Math. Soc. **2**(1953), 288-323.

[5] *On the theory of functions of several complex variables.* II. *A complete orthonormal system in the hyperbolic space of Lie-hypersphere,* Acta Math. Sinica **5**(1955), 1-25. (Chinese. English summary)

[6] *On the theory of functions of several complex variables.* III. *On a complete orthonormal system in the hyperbolic space of symmetric and skew-symmetric matrices,* Acta Math. Sinica **5**(1955), 205-242. (Chinese. English summary)

[7] *Some definite integrals,* Acta Math. Sinica **6**(1956), 302-312. (Chinese. English summary)

[8] *On a system of partial differential equations,* Sci. Record (N.S.) **1**(1957), 369-371.

L. K. Hua, K. H. Look

[1] *On Cauchy formula for the space of skew symmetric matrices of odd order,* Sci. Record (N.S.) **2**(1958), 19-22.

[2] *Boundary properties of the Poisson integral of Lie sphere,* Sci. Record (N.S.) **2**(1958), 77-80.

[3] *Theory of harmonic functions of classical domains.* I. *Harmonic functions in the hyperbolic space of matrices,* Acta Math. Sinica **8**(1958), 531-547. (Chinese. English summary)

J. Mitchell

[1] *The kernel function in the geometry of matrices,* Duke Math. J. **19**(1952), 575-583.

[2] *Potential theory in the geometry of matrices,* Trans. Amer. Math. Soc. **79**(1955), 401-422.

F. D. Murnaghan

[1] *The theory of group representations,* Johns Hopkins Press, Baltimore, Maryland, 1938; Russian transl., IL, Moscow, 1950.

L. S. Pontrjagin

[1] *Continuous groups,* 2nd ed., GITTL, Moscow, 1954. (Russian)

I. M. Ryžik, I. S. Gradšteĭn

[1] *Tables of integrals, sums, series and products,* 3rd ed., GITTL, Moscow, 1951 (Russian); English transl., VEB Deutscher Verlag der Wissenschaften, Berlin, 1957.

C. L. Siegel

[1] *Symplectic geometry,* Amer. J. Math. **65**(1943), 1-86.

G. Szegö

[1] *Orthogonal polynomials,* Amer. Math. Soc. Colloq. Publ. Vol. 23, Amer. Math. Soc., Providence, R. I., 1939.

R. M. Thrall

[1] *On symmetrized Kronecker powers and the structure of the free Lie ring,* Amer. J. Math. **64**(1942), 371-388.

A. Weil

[1] *L'intégration dans les groupes topologiques et ses applications,* Hermann, Paris, 1940; Russian transl., IL, Moscow, 1951.

[2] *L'intégrale de Cauchy et les fonctions de plusieurs variables,* Math. Ann. **111**(1935), 178-182.

H. Weyl

[1] *Harmonics on homogeneous manifolds,* Ann. of Math. (2) **35**(1934), 486-499.

[2] *The classical groups. Their invariants and representations,* Princeton Univ. Press, Princeton, N. J., 1939; Russian transl., IL, Moscow, 1947.

H. Weyl, F. Peter

[1] *Die Vollständigkeit der primitiven Darstellungen einer geschlossenen kontinuierlichen Gruppe,* Math. Ann. **97**(1927), 737-755.

SUPPLEMENTARY BIBLIOGRAPHY

[1] F. A. Berezin, *Laplace operators on semi-simple Lie groups,* Trudy Moskov. Mat. Obšč. **6**(1957), 371-463. (Russian)

[2] F. A. Berezin and I. M. Gel′fand, *Some remarks on the theory of spherical functions on symmetric Riemannian manifolds,* Trudy Moskov. Mat. Obšč. **5**(1956), 311-351. (Russian)

[3] I. M. Gel′fand, *Spherical functions in symmetric Riemann spaces,* Dokl. Akad. Nauk SSSR **70**(1950), 5-8. (Russian)

[4] I. M. Gel′fand and M. I. Graev, *Unitary representations of the real unimodular group (principal nondegenerate series)*, Izv. Akad. Nauk SSSR Ser. Mat. **17**(1953), 189-248. (Russian)

[5] ——, *Geometry of homogeneous spaces, representations of groups in homogeneous spaces and related questions of integral geometry.* I, Trudy Moskov. Mat. Obšč. **8**(1959), 321-390. (Russian)

[6] I. M. Gel′fand and M. A. Naĭmark, *Unitary representations of the classical groups,* Trudy Mat. Inst. Steklov. **36**(1950). (Russian)

[7] I. M. Gel′fand and I. I. Pjateckiĭ-Šapiro, *Theory of representations and theory of automorphic functions,* Uspehi Mat. Nauk **14**(1959), no. 2(86), 171-194. (Russian)

[8] M. I. Graev, *Unitary representations of real simple Lie groups,* Trudy Moskov. Mat. Obšč. **7**(1958), 335-389. (Russian)

[9] C. L. Siegel, *Analytic functions of several complex variables,* Institute for advanced Study, Princeton, N. J., 1950; Russian transl., IL, Moscow, 1954.

[10] F. I. Karpelevič, *Geodesics and harmonic functions on symmetric spaces,* Dokl. Akad. Nauk SSSR **124**(1959), 1199-1202. (Russian)

[11] I. I. Pjateckiĭ-Šapiro, *Discrete subgroups of the group of analytic automorphisms of the polycylinder, and automorphic forms,* Dokl. Akad. Nauk SSSR **124**(1959), 760-763. (Russian)

[12] É. Cartan, *Sur la determination d'un système orthogonal complet dans un espace de Riemann symétrique clos,* Rend. Circ. Mat. Palermo **53**(1929), 217-252.

[13] R. Godement, *A theory of spherical functions.* I, Trans. Amer. Math. Soc. **73**(1952), 496-556.

[14] Harish-Chandra, *Spherical functions on a semisimple Lie group.* I, II, Amer. J. Math. **80**(1958), 241-310, 553-613.

E DUE

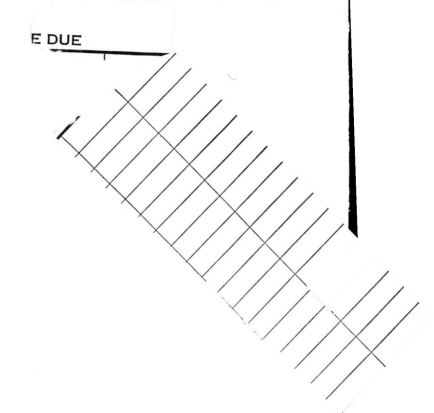